Hunter, Killer

How AI Drone Swarms Are Rewriting the Rules of War

C. Scott Lannon

Published by C. Scott Lannon

ISBN: 978-1-972731-08-6 (Digital) ISBN: 978-1-972731-09-3 (Paperback)

For the humans still in the loop.

Introduction: The $200 War

The operator sat in a basement in Zaporizhzhia, two floors below a residential apartment block that had lost its upper stories to a Russian glide bomb three months earlier. He was twenty-six years old, callsign "Lens," and he had been doing this for nine hours. The room smelled like solder flux, stale coffee, and the particular ozone tang of lithium-polymer batteries on their fourth charge cycle. A commercial monitor, propped on a folding table between coils of antenna cable, showed a grainy green thermal feed from a drone circling at 120 meters above a tree line in the gray zone east of Orikhiv. On Lens's head sat a pair of FPV goggles, first-person view, meaning he saw what the drone saw, the world reduced to a monocular rectangle of thermal contrast and static. His hands rested on a controller that had cost less than a video game console. He had built it from parts ordered online.

The drone itself was not much to look at. Assembled from a commercial racing-drone frame, a repurposed motor, and a guidance board running open-source firmware, it weighed just over a kilogram before the warhead was attached, a modified RPG munition fitted to a 3D printed tail fin, the design shared on a Telegram channel run by a volunteer engineering collective in Dnipro. The entire aircraft, warhead included, cost roughly $200 to produce. The target filling Lens's screen, the hot white bloom of a Russian T-72B3 main battle tank idling near a drainage ditch, its engine ticking over to keep the crew's heater running, was worth approximately $3 million.

Lens thumbed the stick forward. The drone tilted into a dive. On the monitor, the tree line fell away. The tank grew from a bright smear to a defined shape: turret, hull, tracks, the faint glow of exhaust. At three seconds from impact, the tank's outline filled the goggles entirely. Then the feed went black.

In the silence that followed, Lens pulled the goggles to his forehead, blinked in the blue light of the monitor, and reached for the next drone. There were eleven more charged and ready on a plywood rack against the wall. He would fly until the batteries ran out or the sun came up, whichever arrived first.

The Zoom

That basement in Zaporizhzhia is one node in a network so vast, so decentralized, and so consequential that no single government, no defense ministry, and no international body has managed to describe it accurately, let alone control it. In 2024, Ukraine produced over 4 million drones, more than any nation in history, from a manufacturing base of over 500 separate producers, many of them operating at garage scale. Russia matched the effort with 2 million FPV drones and 30,000 long-range strike platforms of its own. Between them, the two countries manufactured more combat drones in a single year than the entire world produced manned military aircraft in the previous decade.

But the drone in Lens's hands was already becoming obsolete, not because it did not work, but because the next generation would not need him.

Three hundred miles west of Zaporizhzhia, in a nondescript office park where a Ukrainian defense-technology startup called

Swarmer had set up its operations center, a team of software engineers was testing the system designed to make the single-operator, single-drone model of war a relic. Their platform, Styx, coordinated reconnaissance and strike drones within autonomous networks, one human setting intent and rules, multiple drones executing without individual pilot control. By September 2025, Styx had logged over 82,000 combat missions. By March 2026, the count exceeded 100,000, the first documented instance of swarm coordination technology used routinely in active combat. That same month, Swarmer filed for an initial public offering on Nasdaq under the ticker SWMR. The stock surged nearly 1,000% in its first three trading sessions. The market's message was unambiguous: combat-proven autonomous weapons were not a future technology. They were a present asset class.

Eight thousand miles from Zaporizhzhia, the scale shifted again. In Irvine, California, Palmer Luckey's Anduril Industries, named for the fictional sword in *The Lord of the Rings*, founded in 2017 with the explicit mission of bringing Silicon Valley's development speed to the Pentagon's acquisition process, had just closed a $4 billion funding round led by Andreessen Horowitz and Thrive Capital. The round valued the company at $60 billion, nearly double the $30.5 billion valuation it had achieved just nine months earlier. The same week, the US Army awarded Anduril a ten-year, $20 billion ceiling contract, the largest single counter-drone agreement in US military history, consolidating more than 120 separate procurement actions into a single enterprise deal. At the center of the contract sat Lattice, Anduril's AI-powered command-and-control platform. Lattice ingests data from distributed sensors (radar, radio frequency, acoustic, autonomous platforms) fuses it using artificial intelligence into a single operational picture, and manages groups of unmanned systems simultaneously across air, ground, and maritime domains. It was designed to do for the entire American military what Swarmer's Styx does for a Ukrainian drone squad: compress the kill chain from detection to destruction into a single, AI-mediated process.

And at the United Nations in New York, on December 2, 2024, the General Assembly had adopted Resolution 79/62, titled "Lethal Autonomous Weapons Systems." The vote: 166 nations in favor, 15 abstentions, 3 opposed. The three nations that voted no were Belarus, North Korea, and Russia. The resolution called for a "comprehensive and inclusive" approach to addressing the challenges posed by weapons that select and engage targets without direct human control, weapons that, as of the vote, did not yet have an agreed legal definition. The UN Secretary-General, the International Committee of the Red Cross, and the Campaign to Stop Killer Robots all called for the conclusion of negotiations by the end of 2026. The technology they were trying to regulate was evolving faster than the diplomatic calendar could accommodate.

The Most Consequential Weapons Revolution Since the Bomb

This book is about what connects the basement in Zaporizhzhia to the conference room in Irvine to the chamber in New York — and about the revolution that is moving faster than any of those institutions can track.

The revolution is not coming. It has already happened. A $200 drone can destroy a $3 million tank. A $200 drone can destroy a $24 million air defense system, as Ukrainian soldiers demonstrated on the Kupiansk front in November 2023, when a civilian-donated FPV drone struck a Russian Tor missile launcher, producing a cost ratio of one to one hundred twenty thousand. An autonomous coordination platform built by a Ukrainian startup can direct over 100,000 combat missions before the company's stock even begins trading. An AI command-and-control system built by a Californian defense firm can manage swarms of unmanned vehicles across every domain of warfare, and the US Army will pay $20 billion for it.

These are not hypotheticals. They are not projections. They are accomplished facts, all occurring within a single eighteen-month window between late 2024 and early 2026. And they represent only the visible edge of a transformation that extends far beyond Ukraine or Silicon Valley. At least 113 countries now operate military drones. Non-state actors from the Houthis to Hezbollah have demonstrated precision strike capability with modified commercial platforms costing less than a used car. China has publicly demonstrated autonomous swarms of over 200 drones in military exercises and is building the manufacturing base to produce them at a scale that dwarfs any Western competitor. The global military drone market, valued at $47.38 billion in 2025, is projected to reach $98.24 billion by 2033. The AI-in-drones market alone is growing at 27.4% annually.

This is the most consequential shift in the technology of killing since the development of the atomic bomb.

The comparison to the bomb is not rhetorical inflation. Nuclear weapons changed warfare by making the consequences of great-power conflict unsurvivable. Autonomous weapons are changing warfare by making the *conduct* of conflict compressible, by collapsing the time between finding a target and destroying it from hours or minutes to seconds or fractions of seconds, and by removing, layer by layer, the human judgment that once occupied that interval. The atomic bomb raised the question of whether war between nuclear states was rational. Autonomous weapons raise a different question, arguably harder and certainly more urgent: as machines take over more of the kill chain, the sequence of detecting, tracking, targeting, and engaging an enemy, at what point does human control become a fiction? And who decides where that point is?

The Hunter-Killer Framework

Every autonomous weapons system in this book, from the $200 FPV drone in Lens's basement to the $60 billion Lattice platform in Anduril's server racks, exists on a spectrum between two functions.

The first is **hunting**: sensing, finding, identifying a target. The reconnaissance drone circling above the tree line near Orikhiv. The Skydio X10D building a 3D map of a contested building at a million points per second. The Lattice platform fusing radar and radio-frequency data into a single operational picture. The Israeli military's Lavender system, which reportedly used machine learning to flag over 37,000 suspected militants in Gaza. These are the hunters: the systems that find.

The second is **killing**: acting on that identification. Striking. Destroying. The FPV drone with the RPG warhead. The Switchblade 600 loitering munition diving onto an armored vehicle. The Baykar K2 kamikaze drone coordinating with four others in an autonomous swarm to suppress and destroy a target without a single command from the ground. These are the killers: the systems that act.

Between those two poles lies a space. Call it **the Gap**. The Gap is where a human being, an operator, a commander, a lawyer reviewing targeting criteria, once made a decision. The reconnaissance drone found the target; a human reviewed the imagery; a different human authorized the strike; a third human guided the weapon. The Gap was filled with judgment, with context, with the capacity to say *no*. With the capacity to distinguish a tank from a school bus, a combatant from a civilian, a legitimate military target from a protected site, not because an algorithm classified it correctly, but because a person looked and understood.

AI is compressing the Gap. Not eliminating it, not yet, not in most systems, not in any publicly acknowledged lethal application. But compressing it, steadily and from both ends. The hunters are getting smarter: drones that navigate without GPS, identify targets without human input, and designate them for strike autonomously. The killers are getting faster: weapons that receive targeting data and act on it in timeframes no human can match. And the systems connecting hunters to killers, the Styxes, the Lattices, the unnamed platforms in Chinese and Israeli and Russian laboratories, are shrinking the interval between detection and destruction toward a vanishing point.

The central question of this book, and the central question of the autonomous weapons revolution, is this: as the Gap compresses, what happens to the human judgment that once lived inside it? Where on the spectrum between hunter and killer does meaningful human control cease to exist, and who gets to decide?

What This Book Will Do

This is not a book about whether killer robots are coming. They are here, in forms more mundane and more consequential than the science-fiction archetype suggests. This is a book about the specific systems, specific people, specific decisions, and specific failures that are shaping how autonomous weapons will be built, deployed, governed, and, inevitably, used against human beings. It is a book for anyone who needs to understand the drone revolution clearly enough to act on that understanding: the military officer writing doctrine, the policy staffer drafting legislation, the engineer building the next targeting algorithm, the investor evaluating a defense-tech startup, the citizen trying to make sense of the drone footage flooding social media feeds from wars they cannot look away from.

The argument proceeds in four parts.

Part I: The Revolution establishes what is happening on the battlefield right now. Ukraine is the proving ground, the largest drone warfare laboratory in history, where 4 million drones produced by 500 manufacturers are rewriting the economics of armored warfare, the tactics of infantry combat, and the meaning of air superiority. The four chapters in Part I trace the full Hunter-Killer spectrum through the Ukrainian crucible and the global ecosystem it has catalyzed: the hunters that find, the killers that strike, the economic inversion that makes a $200 weapon lethal to a million-dollar platform, and the startups and defense primes racing to build the systems that will define the next era.

Part II: The Technology explains how AI drone swarms actually work, not as an abstraction, but as engineering. How a drone navigates without GPS in a jammed environment. How a swarm of twenty aircraft coordinates without centralized control. How a targeting algorithm selects an object from a thermal feed and classifies it as friend, enemy, or civilian. How the counter-drone arms race (electronic warfare, fiber-optic cables, directed energy) drives the push toward greater autonomy by degrading the communications links that human operators depend on. The four chapters in Part II give the reader the technical literacy to evaluate every claim in the debate that follows, without requiring an engineering degree to get there.

Part III: The Stakes confronts what happens when everyone has swarms. Proliferation is not a future risk; it is a present condition. At least 113 countries operate military drones. Non-state actors from the Houthis to Hezbollah have demonstrated precision strike capability using modified commercial platforms. The three chapters in Part III address the ethics of autonomous killing, not the strawman version, but the real debate between those who argue that autonomous weapons reduce casualties and those who argue that removing human judgment from lethal decisions crosses a moral line that no efficiency gain can justify. They examine the great-power competition between the United States and China, where the AI military balance may prove more consequential than the nuclear balance. And they reckon with the question that the UN General

Assembly's 166-nation vote made unavoidable: can international law govern a weapon it cannot define?

Part IV: The Path Forward takes a position. This book is not neutral. The evidence assembled in Parts I through III supports specific conclusions about what the United States must do, in acquisition, in doctrine, in alliance management, and in international norms, to maintain its edge in autonomous systems without ceding the moral authority that distinguishes its use of force from its adversaries'. The three chapters in Part IV offer actionable recommendations for military leaders, policymakers, technologists, and the defense-industrial base. They examine why the Pentagon's Replicator initiative, announced in August 2023 by Deputy Secretary of Defense Kathleen Hicks as the answer to fielding autonomous systems at scale, fell short of its own targets, and what the failure reveals about the structural mismatch between the speed of technology and the pace of military acquisition. They propose a framework for meaningful human control that is neither the full ban demanded by some advocacy organizations nor the unconstrained autonomy that some defense hawks prefer. The honest answer is that the window for shaping this technology's trajectory is open but closing. The path forward requires building faster *and* governing more carefully — simultaneously, not sequentially.

The Promise

By the time you finish this book, you will have a framework, the Hunter-Killer spectrum, that applies to every autonomous weapon, every policy proposal, every defense-tech pitch, and every battlefield development you will encounter for the rest of this decade and beyond. You will understand how a $200 drone destroys a tank, how an AI swarm coordinates without human input, how a targeting algorithm classifies a suspected combatant, and why a $20 billion Pentagon contract matters more than a hundred UN resolutions. You will understand the Gap, the space where human judgment lives or dies, and you will be able to evaluate, for any system or any scenario, whether the humans are still in control or whether the machines have already taken over.

And you will understand why this matters now, not later. The atomic bomb was developed in secret, tested in secret, and used twice before the world grasped what it meant. The autonomous weapons revolution is happening in the open, on Ukrainian battlefields streamed to Telegram, in Silicon Valley pitch decks filed with the SEC, in Congressional hearings broadcast on C-SPAN, in UN votes reported on the evening news. The information is available. What has been missing is the synthesis: a single, clear account that connects the battlefield to the boardroom to the treaty chamber, and that tells the reader not just what is happening but what it means and what must be done.

That is what this book provides.

The drone operator in Zaporizhzhia reached for his next aircraft. In three hundred miles, engineers were building the system that would let one person do what he does twenty times over. In eight thousand miles, a defense contractor was building the platform that would coordinate thousands of those systems across every domain of war. And in New York, 166 nations were trying to write rules for a technology that was already outrunning them.

The revolution is not waiting for permission. Neither can we.

PART I: THE REVOLUTION

Chapter 1: The Proving Ground

Somewhere in Siberia, in the winter of 2025, a nondescript steel shipping container sat in a field ringed by frost-hardened earth. It looked like every other container that had been hauled across Russia by rail and truck for decades, anonymous, industrial, forgettable. Rust streaked its corrugated walls. Snow dusted the roof. From the outside, there was nothing to suggest it contained a weapon.

Inside, a Ukrainian special operations team crouched over commercial monitors and FPV controllers, their breath fogging in the cold. They had smuggled themselves and their equipment across thousands of miles of Russian territory, by methods that remain classified, to position this container within first-person-view drone range of a Russian strategic bomber airfield. Hundreds of miles from the nearest Ukrainian front line, thousands of miles from Kyiv, they were about to attempt the furthest-reaching drone strike of the entire war.

The aircraft parked on the tarmac outside were Tu-95 Bears and Tu-160 Blackjacks, the backbone of Russia's nuclear-capable long-range bomber fleet, aircraft that had been flying deterrence patrols since the Cold War. Each Tu-160 costs an estimated $250 million to build. Russia cannot manufacture replacements at scale. The production lines shut down with the Soviet Union, and while Moscow has attempted a limited Tu-160M restart program, the industrial base that once churned out strategic bombers no longer exists. The specialized tooling, the trained workforce, the supply chains for titanium alloys and avionics: all gone. Every bomber destroyed on that airfield was a permanent subtraction from Russia's strategic arsenal, a hole that could not be patched with money or urgency.

The FPV drones that launched from the shipping container cost a few hundred dollars each.

The operators guided them toward the parked aircraft using first-person-view goggles, the same goggles that teenage drone racers wear at weekend competitions in suburban parks across America, the same controllers sold on Amazon for less than the price of a video game console. The feed from the drone's onboard camera filled the operator's field of vision: gray tarmac, the angular silhouettes of hangars, the massive swept wings of bombers that had once carried nuclear cruise missiles across the Arctic. The drone closed the distance. The target filled the screen. Impact. The feed went to static. The operator set down the controller and reached for the next one. There were more drones charged and ready.

Operation Spider, as it came to be known, was not just the furthest-reaching drone strike of the war. It was the most vivid illustration of a reality that every military planner on Earth was scrambling to understand: a weapon that costs less than a decent bicycle can now impose strategic damage on a nuclear-armed superpower. Not tactical damage—a destroyed truck, a disabled radar post. Strategic damage. The permanent destruction of irreplaceable assets that underpin a nation's nuclear deterrent posture.

Consider what had just happened. A team operating from a steel box in a field had used commercially available components (motors, cameras, batteries, flight controllers, all purchasable online) to destroy aircraft that represented decades of Soviet and Russian aerospace engineering, billions of dollars of accumulated investment, and a pillar of Russia's strategic nuclear triad. The operators did not need satellite-guided cruise missiles, stealth aircraft, or the backing of a superpower's industrial base. They needed a shipping container, an internet connection for parts procurement, and the skills to fly a racing drone. The era in which only great powers could threaten great powers was over. The container in Siberia proved it.

The World's Largest Drone Laboratory

Ukraine did not set out to become the proving ground for the most significant weapons revolution since the atomic bomb. It was forced into it.

When Russia launched its full-scale invasion in February 2022, Ukraine's military was outgunned in nearly every conventional category, tanks, artillery, aircraft, naval power, and sheer troop numbers. What Ukraine had, in modest quantities, was a fleet of Turkish made Bayraktar TB2 drones, purchased before the invasion at a unit cost of roughly $5 million apiece, about 20 aircraft in the entire fleet, according to a spokesperson for Ukraine's air force command. And it had something no procurement office could have planned for: a population of technically skilled civilians, engineers, software developers, hobbyists, drone racing enthusiasts, who understood that survival would require improvisation at a speed and scale that no military doctrine had anticipated.

The TB2 became the first star of the drone war. In the chaotic opening weeks of the invasion, when Russian forces advanced in disorganized columns without establishing proper air defenses, Ukrainian TB2s devastated targets of opportunity. They struck command posts, tanks, supply trucks, fuel trains, patrol boats, and surface-to-air missile systems, including Russia's own Buk and Tor air defense launchers, the very platforms designed to shoot drones down. On May 7, 2022, in one of the war's most cinematic moments, a TB2 destroyed a Russian Mi-8 transport helicopter on Snake Island at the exact moment it was unloading passengers. The same sortie destroyed a Tor missile launcher on the island and a second Tor being offloaded from a landing craft, clearing the way for a Ukrainian Su-27 bombing run on the island's remaining Russian positions. The footage went viral within hours. A folk song written in the Bayraktar's honor became a cultural phenomenon across Ukraine, a dark, jaunty anthem to the drone that was humiliating one of the world's largest armies.

The TB2's dominance lasted less than four months.

By summer 2022, Russian forces had established extensive electronic warfare (EW) infrastructure across the Donbas, jamming complexes deployed at concentrations of up to 10 systems per 21 kilometers of frontage. These systems targeted GPS signals, disrupted the radio links between drones and their operators, and spoofed navigation data to send drones spiraling off course. The electronic barrage was dense, layered, and devastatingly effective. Small quadcopter drones had an average life expectancy of three flights. The TB2, larger and slower, averaged six flights before being jammed, spoofed, or shot down. By late 2022, electronic warfare had contributed to the destruction of 90% of the thousands of drones Ukraine possessed at the start of the invasion, according to an analysis by the Royal United Services Institute (RUSI). Only one-third of Ukrainian drone missions could be considered successful.

The TB2, the folk hero, was being grounded by physics.

Colonel Volodymyr Valiukh of the Ukrainian Main Intelligence Directorate put it bluntly in late October 2023: "For the TB2, I do not want to use the word useless, but it is hard to find situations where to use them." The remaining TB2s were pulled back from strike missions and relegated to short-duration reconnaissance flights at the edge of Russian air defense range, staying out of harm's way while using their optics to guide other weapons. Five million dollars apiece, and they had been reduced to expensive binoculars.

The TB2 story could have been a cautionary tale about drone hype meeting battlefield reality, proof that electronic warfare would always defeat the flying robots. Instead, it became the prologue to an arms race that would transform the conflict, and the future of warfare, beyond recognition.

The $200 Inversion

What happened next was not a technology breakthrough. It was an economics breakthrough.

As Turkish-made drones became too expensive and too vulnerable to risk over the front lines, Ukraine's decentralized network of engineers, hobbyists, veterans, and volunteer groups began assembling something cheaper, faster, and far more disposable: FPV drones, first-person view, meaning the operator sees through the drone's camera via goggles, built from commercial DJI components, 3D-printed tail fins, and repurposed RPG warheads. The unit cost started around $400 and plummeted toward $200 as production scaled and supply chains matured. Two hundred dollars. The price of a pair of running shoes.

The logic was not subtle: if Russian electronic warfare was going to destroy most of your drones before they reached their targets, the answer was not to build better drones. It was to build so many that the losses did not matter. Quantity became the strategy. Disposability became the design philosophy. Every drone was built to be lost.

This was a conceptual inversion that military procurement officers in Washington, London, and Berlin struggled to absorb. Western defense culture had spent decades optimizing for capability per unit, making each platform more survivable, more lethal, more expensive. The Ukrainian model optimized for lethality per dollar. An FPV drone did not need to survive. It did not need to return. It did not need maintenance schedules, spare parts inventories, or trained depot-level technicians. It needed to fly once, hit once, and die. The entire platform was the munition.

The numbers tell the story of what that philosophy produced. By the end of 2024, Ukraine had manufactured more than 4 million drones. Not four thousand. Four million. The manufacturing base was not a single factory or a government arsenal. It was a distributed ecosystem of more than 500 producers, many operating at garage scale, workshops in basements and repurposed warehouses, small teams turning out a few hundred units a month, larger operations producing thousands, all feeding a front line that consumed drones the way the Western Front consumed artillery shells in 1916. Production capacity was projected to reach 7 to 10 million units, aerial, ground, and maritime: by 2026. A single country, under invasion, with a GDP roughly one-tenth of Russia's, was outproducing its adversary in the one weapons category that mattered most on the modern battlefield. Ukraine's Minister of Digital Transformation, Mykhailo Fedorov, had made drone production a national priority, treating it not as a military procurement problem but as a technology mobilization challenge, the kind of effort that a country with a large software industry and a culture of entrepreneurial improvisation could execute faster than any centralized defense ministry. The result was an ecosystem that looked less like a defense industrial base and more like a tech startup scene, except the product was lethal and the customers were soldiers.

Russia, for its part, was not standing still. Moscow's own production targets for 2025 called for 2 million FPV drones and 30,000 long-range strike drones. The Lancet, a loitering munition manufactured by the Kalashnikov Group's ZALA Aero subsidiary, became Russia's signature precision drone weapon, a double-X-winged guided munition that could loiter over an area and strike targets of opportunity. The Lancet forced Ukrainian artillery crews to weld chain-link cages and wire mesh over their howitzers, building improvised shelters that looked medieval but worked often enough to matter. Ukrainian forces parked inflatable HIMARS decoys next to real launchers, hoping to waste Lancets on rubber. Russia was learning, adapting, building. This was not a one-sided revolution. It was an industrial arms race at a tempo not seen since the Second World War.

But the cost math was devastating, and it cut in only one direction.

On November 9, 2023, on the Kupiansk front in eastern Ukraine, Ukrainian soldiers used a civilian-donated FPV drone to destroy a Russian 9K330 Tor surface-to-air missile system. The Tor is one of Russia's most sophisticated short-range air defense platforms, a tracked vehicle carrying eight missiles and an integrated radar, designed to protect ground forces from aircraft and precision munitions. It costs approximately $24 million to build. The drone that killed it cost a few hundred dollars. It had been donated by a volunteer group, civilians pooling money online, buying the components from commercial suppliers, and shipping the assembled weapon to the front in a package no larger than a shoebox.

The math: $24 million. One Tor missile system. That money could have purchased 120,000 of the drones that destroyed it.

This was not an anomaly. It was the new arithmetic of the battlefield. Across hundreds of kilometers of front line, $200 weapons were routinely destroying equipment worth tens of thousands to tens of millions of dollars. Tanks. Armored personnel carriers. Self-propelled artillery. Air defense systems. Electronic warfare stations. Supply trucks. Ammunition depots. The footage accumulated on Telegram channels by the tens of thousands of clips, drone's-eye-view video of a cheap quadcopter diving into the engine deck of a tank, the thermal flash, the secondary explosion as ammunition cooked off. Each clip was a data point in an argument that no amount of doctrinal resistance could refute. The entire logic of conventional armored warfare, which for a century had assumed that expensive, heavily protected platforms could survive through combinations of armor, speed, and firepower, was being rewritten by a drone that an engineering student could assemble in an afternoon from parts ordered on the internet.

10,000 Drones a Month

The scale of attrition was staggering, and it revealed a new kind of warfare, one measured not in territory gained but in platforms consumed.

As of mid-2023, Ukraine was losing between 5,000 and 10,000 drones every month, roughly 160 to 300 per day, according to Ukrainian soldiers and a May 2023 report by RUSI. Most were destroyed by Russian electronic warfare before they reached their targets. The jammers worked. The drones died. And Ukraine built more. The production lines ran day and night, every workshop in the ecosystem pushing units to the front as fast as components could be sourced, assembled, tested, and shipped.

This cycle, jam, lose, replace, repeat, defined the character of the war from mid-2022 onward. It was a war of industrial attrition waged not with steel and gunpowder but with circuit boards and lithium-polymer batteries, and it demanded a production tempo that would have been unimaginable to any defense ministry two years earlier. The side that could produce drones faster, cheaper, and in greater volume gained the advantage. The side that could jam, intercept, or destroy those drones denied it. Every innovation on one side provoked a countermove on the other, and the cycle accelerated with each passing month. By 2024, both sides combined were fielding more drones per month than the entire global drone industry had produced in a year before the war. The consumption was relentless. Drones were no longer weapons in the traditional sense, they were ammunition, consumed by the thousands and replenished by the tens of thousands, each one a single-use precision munition with a shelf life measured in minutes from takeoff to impact or destruction.

Russia's electronic warfare campaign was not a sideshow. It was one of the most effective defensive operations of the entire war. EW posts were established approximately every 10 kilometers along the entire length of the front line, positioned about 6 kilometers behind the forward edge of the battle area. The Palantin system, among Russia's most capable EW platforms, could suppress satellite radio navigation signals along the entire line of contact and replace them with spoofed coordinates, GPS data that told a drone it was somewhere it was not, sending it spiraling into the ground or drifting harmlessly off target. Russia also deployed the Borisoglebsk-2, with its wider frequency bandwidth and faster spectrum scanning, and the Layer system for broader area suppression. By October 2023, *The Economist* reported that electronic warfare was in widespread use across the front, with Russia installing video feedback and control jammers directly onto high-value equipment, tanks, artillery pieces, command vehicles, turning each platform into its own electronic shield.

The density of this electronic defense was unprecedented in modern warfare. Russian forces built jammers small enough to fit in a backpack. Pocket-sized jammers were issued to individual soldiers. Every Ukrainian drone that flew toward the front line entered a gauntlet of electronic fire that began kilometers before it reached its target. The airspace above no-man's-land was not empty. It was saturated with invisible energy, a wall of electromagnetic interference designed to make every frequency unusable, every GPS coordinate a lie.

The jammers were winning. The drones kept dying.

And then the Ukrainians adapted.

The Wire and the Ghost

The adaptation came in two forms, and both pointed toward the same future: a battlefield where human control of individual drones becomes optional.

The first adaptation was physical, an old idea resurrected for a new war. In October 2024, a spokesperson for a Ukrainian brigade operating in Zaporizhzhia Oblast reported that Russian forces had begun deploying FPV drones trailing fiber-optic cables stretching up to 10 kilometers behind them. The concept was borrowed from a weapon system that most military planners had assumed was a relic: the wire-guided missile, the kind of technology that Cold War tank crews trained to use before laser and GPS guidance made it obsolete. A thin fiber-optic cable, spooled out behind the drone as it flew, carried the control signals and video feed through a physical wire rather than a radio frequency. No radio signal meant no signal to jam. No GPS link meant no link to spoof. No electromagnetic emission meant the operator's position stayed hidden. The drone was, in the precise technical sense, unjammable.

Ukrainian personnel confronting these drones for the first time had no electronic countermeasure that worked against them. The jammers that had been destroying thousands of drones per month were suddenly useless against this variant. Desperate soldiers resorted to shooting fiber-optic drones out of the sky with rifles and machine guns, small arms against a target the size of a dinner plate, moving at 80 kilometers per hour, trailing a nearly invisible thread of glass. It was a scene that would have been familiar to an anti-aircraft gunner in 1944.

Both sides adopted fiber-optic drones with startling speed. The cables ranged from 5 to 20 kilometers in length. The drones could idle on the ground for hours in ambush positions, consuming almost no power while they waited for a target to pass within range. Their control latency was lower than radio-linked drones, their video quality higher, their survivability dramatically better. They could operate in locations where radio contact was poor or impossible, urban canyons, deep valleys, areas saturated with jamming. The tradeoffs were real: reduced maneuverability, the physical constraint of the cable, the risk of snagging on terrain. But in practice, the reduced latency and increased survivability often made fiber-optic drones more effective than their wireless counterparts, not less. By May 2025, *The Washington Post* was reporting that Ukraine was scrambling to overcome Russia's growing edge in fiber-optic drone technology, a remarkable reversal for a country that had been the clear innovator in drone warfare only months earlier.

That was the physical adaptation. The second was algorithmic, and its implications ran deeper.

In March 2024, footage posted on social media showed a Ukrainian FPV drone approaching a Russian position when its video feed broke apart, the telltale visual signature of an electronic warfare system severing the radio link between the drone and its operator. The operator lost control. The drone should have tumbled out of the sky or drifted off course, another casualty of Russian jamming. Instead, it completed the strike. An onboard algorithm, trained to recognize the target's visual signature and execute the terminal attack phase without human input, guided the drone through the final seconds of its flight and into the target. The operator's screen was static. The algorithm finished the job.

This was automated terminal attack capability: the moment when a drone no longer needs its operator to complete the kill. The human sets the course, identifies the target, guides the drone through the approach. But in the critical final seconds, the seconds where electronic warfare is most intense, where the jammer has the best chance of severing the link between pilot and weapon, the machine takes over. It locks onto the visual profile of the target, adjusts its flight path, and drives itself to impact without any human guidance. It was not full autonomy, not a machine choosing its own targets, not a system operating without human oversight from start to finish. The operator remained in the loop for target selection and the majority of the flight. But in the gap between human control and detonation, a gap measured in seconds, a gap created by the enemy's jammer, an algorithm had filled the space where a person used to be. The human chose the target. The machine completed the kill.

The implications extended far beyond a single drone strike in eastern Ukraine. The automated terminal attack was a crack in a wall, and the logic of warfare was pushing through it. If the terminal phase could be automated, the approach phase could follow. If the approach phase, then navigation. If navigation, then target identification itself. Each step compressed the Hunter Killer spectrum, the gap between finding a target and striking it, into an ever-smaller window of human involvement. And with each compression, the space for a human hand on the controls shrank. The drone that completed a strike after its operator was jammed out of the loop was not fully autonomous. But it was autonomous enough. And "autonomous enough" is where the revolution lives.

The Startup That Went to War

The company that took this logic furthest (the company that built a business on the premise that the future of drone warfare belongs to software, not hardware) was not a government program or a legacy defense contractor with decades of Pentagon relationships and lobbyists on K Street. It was a Ukrainian startup called Swarmer, born in a war zone, funded on the premise that combat data is the most valuable asset in the defense-tech world.

Swarmer's flagship product, the Styx system, does not fly drones. It coordinates them. Styx operates as an autonomous network layer that sits between the human commander and the drone fleet, managing the interaction between reconnaissance drones that find targets and strike drones that hit them. Groups of drones communicate with each other, share targeting data, adjust their positions, and adapt to battlefield conditions, all without requiring a human to pilot each aircraft individually. The human sets the intent: the area to search, the target type to prioritize, the rules of engagement. Styx handles the coordination. One operator, many drones. Intent-based control rather than stick-and-rudder piloting.

The distinction matters, because it represents a fundamental shift in how humans relate to weapons. A human flying an FPV drone through goggles is a remote-control pilot, skilled, brave, but limited to controlling one aircraft at a time, burning through attention and adrenaline on every run. The operator's hands shake. The operator's eyes fatigue. The operator can fly one drone, then another, then another, but never two simultaneously. Styx changed the equation. It turned operators into commanders. Instead of steering a single drone through a single attack run, a Styx operator defined the mission parameters and let the software coordinate the swarm. Reconnaissance drones identified targets and passed coordinates to strike drones. Strike drones adjusted approach vectors based on the reconnaissance data. If one drone was jammed or destroyed, the others adapted without waiting for new human instructions. The system did not think, it executed coordination protocols, following rules its designers had written. But the effect, from the target's perspective, was indistinguishable from a coordinated attack by multiple intelligent adversaries approaching from different directions simultaneously.

By September 2025, Swarmer's technology had logged more than 82,000 combat missions on the front lines of the Ukraine war. Not simulations. Not exercises. Not demonstrations at defense trade shows where executives clap politely over catered lunches. Combat missions, under fire, against an adversary operating some of the most sophisticated electronic warfare systems on Earth. By March 2026, when Swarmer filed for its initial public offering, the mission count exceeded 100,000.

Ukraine had also tested swarm operations at larger scales—more than 100 operations involving groups of 8 to 25 drones operating as coordinated units. Military analysts at multiple institutions identified these as the first documented cases of swarm technology used routinely in active combat—not a single dramatic demonstration, but a sustained operational practice integrated into the daily rhythm of the war. The model pointed toward a near-term future where the fundamental ratio shifted: not one operator per drone, but one operator per swarm. Not piloting, but commanding. Set the rules. Define the mission. Let the software execute.

In March 2026, Swarmer listed on the Nasdaq exchange under the ticker SWMR. In its first three trading sessions, the stock surged nearly 1,000%.

Wall Street was pricing in something that defense ministries around the world were still debating in committee rooms and policy papers: that combat-proven autonomous coordination software, validated under the most punishing conditions on Earth, was worth a fortune. Swarmer had not just built a product. It had built a combat record that no competitor anywhere could match, 100,000 missions of real-world data on how autonomous drone networks behave under fire, how they adapt to electronic warfare, how they coordinate strikes across a contested battlespace. That data was the company's moat, deeper than any patent portfolio or government contract. The financial markets understood the value of 100,000 combat data points before the Pentagon's acquisition system could write a requirements document for a comparable capability.

The Laboratory's Verdict

The story of drones in Ukraine is not a story of one side's triumph. It is a story of relentless, grinding adaptation, each innovation met by a countermove, each countermove spawning the next evolutionary cycle.

Russia deployed electronic warfare at a density and effectiveness no military had previously achieved in combat. Ukraine responded with volume, millions of cheap drones that could absorb catastrophic losses and keep attacking. Russia adapted its EW systems to target new frequency bands and control protocols. Ukraine developed fiber-optic drones that bypassed electronic warfare entirely. Russia built its own.

Ukraine deployed automated terminal attack software that allowed jammed drones to complete their strikes without human control. Russia bolted physical countermeasures onto everything that moved, cage armor welded onto tanks, anti-drone grilles on submarines, video jammers mounted directly on artillery pieces. Ukraine pushed toward swarm coordination to overwhelm point defenses with numbers no human operator could individually manage.

Each cycle compressed the timeline. The gap between a new tactic and its countermeasure, which took months in 2022, shrank to weeks by 2024 and days by 2025. The front line became a laboratory operating at the speed of software updates, not hardware procurement cycles. A Ukrainian workshop could observe a Russian countermeasure on Monday, redesign a drone's guidance firmware by Wednesday, and push the update to operators on the front by Friday. Russia's defense industry, for all its bureaucratic weight, proved capable of similar speed in its strongest domains, electronic warfare and loitering munitions. Neither side could hold an advantage for long. The only constant was the acceleration itself.

This is the dynamic that separates Ukraine from every previous conflict involving drones. In Nagorno-Karabakh in 2020, Azerbaijan's Turkish and Israeli-made drones annihilated Armenian air defenses in 72 hours, but Armenia never adapted. The war was too short and the technological gap too wide. In Ukraine, both sides possessed the engineering talent, the industrial capacity, and the motivation to innovate under fire. The result was not a one-sided demonstration but a genuine arms race, the first full-speed drone arms race in history, fought not between defense ministries over decades but between workshop teams over days.

This pace of adaptation (this relentless, grinding, iterative cycle of innovation under fire) is what makes Ukraine the most important military conflict since 1991. The Gulf War demonstrated that precision munitions could annihilate a conventional army in days. Ukraine is demonstrating something more unsettling: that the tools of precision warfare have been democratized to the point where a volunteer group can crowdfund a weapon that destroys a $24 million air defense system, where a startup with 100,000 combat missions can list on Nasdaq, where a shipping container in Siberia can threaten a nuclear power's strategic bomber fleet.

The lessons are not staying in Ukraine. They cannot. The footage is on Telegram and YouTube. The tactics are being dissected in war colleges from Annapolis to Beijing. The production techniques are being reverse-engineered by every country with an electronics industry and a reason to worry about its neighbors. In January 2026, the Pentagon launched the Orchestrator Prize Challenge, a $100 million competition for AI software that can command drone swarms across air, ground, and maritime domains using voice and text commands. SpaceX and xAI are among the participants. In Sweden, Saab demonstrated a system allowing a single operator to control 100 drones simultaneously, developed in 12 months rather than the typical five-year procurement cycle. In Norway, the military fielded the first operational NATO drone swarm in October 2025. In Ukraine itself, the K2 Brigade stood up the world's first unmanned ground vehicle battalion, and domestic manufacturer Tencore built more than 2,000 UGVs in 2025 with orders for approximately 40,000 units in 2026.

The global military drone market, valued at $47.38 billion in 2025, is projected to reach $98.24 billion by 2033. The

AI-in-drones segment is growing faster still, from $821 million in 2025 to a projected $2.75 billion by 2030. The counter-drone market already exceeds $4 billion. The FY2026 National Defense Authorization Act (NDAA) alone allocated $3.187 billion for counter-unmanned aircraft systems (C-UAS).

The money is a signal. Governments do not allocate billions for fads. They allocate billions when the evidence becomes impossible to ignore, and the evidence from Ukraine is impossible to ignore. Cheap, disposable, increasingly autonomous drones have permanently altered the economics of warfare, the survivability of conventional platforms, and the role of the human operator. What began in Ukrainian basements and garages, improvised, volunteer-funded, assembled from commercial parts, has become the most consequential military innovation of the twenty-first century. The proving ground has delivered its verdict. Every military that wants to remain relevant is racing to absorb it.

But the FPV drone that destroyed the Tor missile system is only the killing half of the equation. Before anything can be killed, it must first be found. And finding targets, persistently, autonomously, in GPS-denied and electronically contested environments, turns out to be the harder problem. An American company founded by three MIT roboticists thinks it has the answer.

Chapter 2: The Hunters

PART I: THE REVOLUTION

On a Tuesday morning in March 2026, Adam Bry stood on the production floor of Skydio's 70,000-square-foot manufacturing facility in Hayward, California, and watched drones come off the line. Each X10D moved through 550 individual quality checkpoints before a technician cleared it for shipment. A single unit assembled in approximately nine minutes. The facility was producing more than a thousand per month, and Bry needed it to produce more. Three days earlier, the US Army had placed a $52 million order for over 2,500 X10Ds, the largest single-vendor tactical small unmanned aircraft system (sUAS) order in Army history. The contract had moved from bid to award in under 72 hours, bypassing the multi-year acquisition timelines that typically define Pentagon procurement. Seventy-two hours. In an institution where it can take longer to requisition office furniture, the Army had bought 2,500 autonomous drones in the time it takes most government offices to process a travel voucher.

Bry picked up an X10D from the line. It was compact enough to deploy from a backpack in 40 seconds; folded, it measured roughly 14 inches long by 6 inches wide. Unfolded, it looked like a consumer quadcopter, the kind a real estate agent might use to photograph a rooftop. It was not a consumer quadcopter. Bry pointed to the six navigation cameras arranged around its body, three on top, three on bottom, that give the drone 360-degree awareness with no blind spots. "DJI drones are remote-controlled aircraft with obstacle sensors added on," he said. "This is an autonomous robot that happens to fly."

The distinction is not marketing. It is the engineering architecture that 29 allied nations now depend on. And it represents the less visible but more consequential half of the drone revolution, the half that does not involve explosions, viral combat footage, or $200 weapons destroying million-dollar tanks. Before anything can be killed, it must first be found. The ability to find it, persistently, autonomously, in GPS-denied and electronically contested environments where satellite signals are jammed and communications are degraded, is the capability that separates a battlefield nuisance from a strategic weapon system.

This is the story of the hunters.

From MIT to the Sky

Skydio's origins bear no resemblance to a defense contractor. The company traces back to 2009, to a robotics lab at the Massachusetts Institute of Technology, where two graduate students, Adam Bry and Abraham "Abe" Bachrach, converged in the Computer Science and Artificial Intelligence Laboratory (CSAIL) on a problem that seemed, at the time, purely academic: how to make an aircraft fly itself without GPS.

Bry was pursuing his master's degree in Aeronautics and Astronautics. Bachrach, who had arrived from UC Berkeley with an undergraduate degree in Electrical Engineering and Computer Sciences, was working on his PhD under Professor Nicholas Roy. His dissertation, "Trajectory Bundle Estimation for Perception-Driven Planning," was a mouthful that described a deceptively simple idea, giving a flying machine the ability to perceive its environment, build a map of that environment in real time, and plan a safe path through it, all without relying on any external signal. The two of them built a fixed-wing drone that could navigate autonomously through a parking garage using nothing but a laser range finder and onboard computation. No GPS. No human pilot. The drone saw the world and figured out where to go.

In 2012, Bry and Bachrach joined the founding team of Google X's Project Wing, an ambitious and well-funded effort to develop delivery drones. They spent 18 months in Mountain View building autonomous-control algorithms and testing delivery drones in Australia. The experience taught them something they had not expected: the future of drones was not in delivering packages. The hard problem, the problem worth a company, was autonomous intelligence itself. The ability to see, understand, and navigate the physical world without human guidance or satellite signals. Everything else was an application built on top of that core capability.

They recruited Matt Donahoe, a fellow MIT graduate with deep software experience, and on April 8, 2014, founded Skydio in Redwood City, California. The founding thesis was straightforward: build drones that can truly fly themselves using AI and computer vision, not drones that require a skilled human pilot wrestling with joysticks. The name itself, Skydio, was a compression of intelligence in the sky. Andreessen Horowitz, the venture capital firm that would later lead the "American Dynamism" thesis investing explicitly in companies that strengthen national security, put in early money. Accel led a $3 million seed round. The clock started.

What Bry, Bachrach, and Donahoe did not know in 2014 was that the technology they were building for consumers, for hikers who wanted a drone that could follow them through the trees, for real estate agents who wanted autonomous site surveys, would become, within a decade, the foundational sensing capability for the most significant military drone deployment in NATO history.

A Drone That Sees for Itself

The Skydio Autonomy Engine is the technical core of everything the company has built, and understanding it is essential to understanding why 29 nations chose this particular drone for their militaries. The engine emerged directly from Bachrach's PhD research at MIT. It was not adapted from existing drone software or licensed from a third party. It was built from the ground up over more than four years of research and development, and it represents what is arguably the most advanced visual-navigation AI for small drones in the world.

The system works like this. The X10D's six navigation cameras capture the world around the drone in every direction, no blind spots above, below, or to any side. That visual data feeds into a processing pipeline running on an NVIDIA Jetson Orin GPU, a chip roughly the size of a credit card that provides the kind of computing power that would have required a server rack a decade ago. The GPU runs nine custom deep neural networks simultaneously, each handling a different aspect of perception and planning. One network estimates the drone's own motion through space. Another builds a three-dimensional obstacle map at more than one million points per second, a constantly updating model of every wall, branch, cable, and surface within the drone's visual range. A third network predicts the path of any subject the drone is tracking, whether a person walking or a vehicle driving. A fourth plans collision-free trajectories through the mapped environment. A fifth handles the aerodynamic calculations for aggressive maneuvering. All of this happens in real time, multiple times per second, onboard the drone itself, not on a remote server, not in a cloud, not through a data link that an adversary could jam.

The critical capability for military operations is a technology called Visual-Inertial Odometry, or VIO. This is Skydio's solution for GPS-denied environments, the battlefield reality of 2026, where Russian electronic warfare units have demonstrated they can jam, spoof, or degrade satellite navigation signals across hundreds of square kilometers. VIO fuses computer vision data from the navigation cameras with inputs from the drone's inertial measurement unit, an onboard sensor that tracks acceleration and rotation, to determine the drone's position and orientation without any external signal. The drone holds position, navigates complex environments, and returns home even when every satellite signal in the sky has been neutralized. On the X10D, VIO operates reliably at altitudes up to 300 meters.

The underlying framework is SLAM, Simultaneous Localization and Mapping, the same class of technology that makes self-driving cars work, but adapted for three-dimensional aerial navigation instead of two-dimensional roads. The drone builds a map of its environment while simultaneously tracking its own position within that map, updating both continuously. It is, in the most literal sense, a robot that understands where it is and what surrounds it.

The practical difference between this approach and DJI's approach is not incremental. It is architectural. DJI, the Chinese company that controls more than 70% of the global consumer drone market, builds remote-controlled aircraft with obstacle sensors added on. The sensors help, but they have significant blind spots. They do not predict motion. They cannot infer that a cable extends beyond what the camera can see. They depend on GPS for basic navigation. When a DJI drone encounters an obstacle while tracking a subject at speed, it slows down or stops. In the DJI Mavic's high-speed Sport Mode, obstacle avoidance is disabled entirely. Skydio's architecture assumes the opposite premise: the AI is the primary pilot, and the human provides intent, judgment, and mission priorities. The drone handles everything else.

The Consumer Graveyard

Skydio did not arrive at the defense market by design. It arrived there by failure.

The company's first product, the Skydio R1, launched in February 2018, a 12-camera drone with impressive autonomous tracking capabilities and a price tag of $2,499. It was a proof of concept that established Skydio's reputation among tech reviewers but sold in limited numbers. The Skydio 2, released in 2019 at a much lower price point of roughly $1,000, was the company's attempt at a consumer breakthrough. It was, by most accounts, a remarkable piece of engineering, smaller, faster, and more capable than the R1, with six 4K navigation cameras and the kind of autonomous obstacle avoidance that made DJI's offerings look primitive by comparison.

The problem was that most consumers did not care about autonomous obstacle avoidance. They cared about price, image quality, and availability. And on all three counts, DJI was untouchable. The Skydio 2 cost roughly twice as much as the comparable DJI Mavic, and DJI's image quality, stabilization, and camera features were superior for the photography and videography work that most consumer buyers actually wanted. Skydio's wait times stretched to months. The company was losing money on every consumer order. By August 2023, Skydio made the decision that would define its future: it exited the consumer drone market entirely, joining what one industry observer called "an ever-growing list of American drone companies that have tried to build camera drones under $1,000, and failed."

The failure was clarifying. If Skydio could not beat DJI on price (and no American manufacturer could, given the cost advantages of Chinese manufacturing) then the company needed customers who valued something DJI could not offer. GPS-denied autonomous navigation. Electronic warfare resilience. Secure data architecture. Compliance with the National Defense Authorization Act. The customers who needed those things were not hikers and real estate agents. They were soldiers.

The Defense Pivot

In July 2020, Skydio launched the X2, its first platform designed explicitly for enterprise and defense customers. The X2D, the defense variant, added thermal imaging from FLIR, a folding design for field portability, and the secure data architecture that government buyers required. It was the drone that caught the Pentagon's attention. In February 2022, the US Army selected Skydio as the sole platform for Tranche 1 of the Short-Range Reconnaissance (SRR) Program of Record, the Army's effort to field small reconnaissance drones to platoon-level units. The contract was worth up to $99.8 million over five years.

The SRR win was transformative. For the first time, an autonomous drone built by a venture-backed Silicon Valley startup (not a legacy defense contractor like Northrop Grumman or General Atomics) had become a program of record for the US Army. The implications rippled through the defense-tech ecosystem. If Skydio could win a program of record, so could Shield AI, so could Anduril, so could any startup with superior technology and the willingness to work through Pentagon procurement.

Then came Ukraine. Starting in 2022, Skydio delivered approximately 1,000 drones to the Ukrainian Defense Forces. Skydio teams made more than 30 visits to the combat zone over two years, an extraordinary level of engagement for an American company in an active war. The drones were linked to Ukraine's Delta battlefield management system, operating as the "hunter" half of a pairing that was becoming standard practice: a Skydio X10D locates the target, and a killer drone, an FPV or a loitering munition: destroys it. The feedback from Ukrainian operators flowed directly to Skydio's engineers in California, sharpening the X10D's design with lessons that no testing range could replicate. "I have never met drone users as sophisticated as the folks in Ukraine," Bry said.

The war validated something Bry had been arguing since Skydio's founding: autonomy was not a feature. It was the entire point. In the electronic warfare environment of eastern Ukraine, where Russian jamming units could knock a GPS-dependent drone out of the sky in seconds, a drone that navigated by vision (that did not need satellites to know where it was or where it was going) was the difference between a useful intelligence asset and an expensive piece of debris. Ukrainian operators were not buying Skydio drones because they were American-made. They were buying them because they worked when nothing else did.

By late 2024, more than half of Skydio's business came from military customers. Revenue hit an estimated $180 million, up 80% year over year. The bookings pipeline reached $1.2 billion, with over half from defense contracts. The company's valuation climbed past $2.2 billion. In August 2025, the NATO Support and Procurement Agency (NSPA) selected Skydio for a framework agreement covering nano-category drone procurement and lifecycle support across member nations. Fifteen NATO countries signed on immediately. The total grew to 29 allied nations using Skydio systems, every branch of the US military, every member of the Five Eyes intelligence alliance, and militaries from Norway to India to Australia.

The Cracks in the Story

Skydio's trajectory from MIT lab to NATO standard is compelling, but the full picture includes setbacks that its marketing materials do not advertise.

The most significant was the SRR Tranche 2 loss. In November 2024, the Army awarded the SRR Tranche 2 Program of Record contract not to Skydio but to Red Cat Holdings, the parent company of Teal Drones. The Black Widow drone that Red Cat offered was smaller, lighter, and, critically, cheaper. The

Army's stated rationale was competition: fostering multiple vendors rather than locking into a single supplier. Skydio continued to receive SRR orders alongside Red Cat, winning a $7.9 million low-rate initial production contract in partnership with SAIC in October 2025, but losing the primary program of record was a competitive wound that the company's critics seized on.

Then there was the supply chain crisis. Despite marketing itself as "the first and only company manufacturing small autonomous drones at scale in the U.S.," Skydio historically relied on Chinese components, including, critically, lithium-ion batteries from Dongguan Poweramp, a TDK subsidiary. In October 2024, China imposed sanctions that blocked Skydio's access to its sole battery supplier. The company was forced to ration batteries to one per drone. In March 2025, China placed Skydio and 14 other US entities on its export control list. The irony was painful: the company positioning itself as the American alternative to Chinese drones could not build its product without Chinese batteries. Skydio scrambled to secure domestic suppliers and by early 2026 was touting a "fully domestic" supply chain, but the transition exposed a vulnerability that critics had warned about for years.

And the price gap persisted. The X10D cost approximately $17,300 per unit, based on the implied pricing of the March 2026 Army order. A comparable DJI Mavic 3 cost roughly $5,000. That three-to-four-times price premium was the single biggest barrier to replacing DJI at scale, and it fueled a heated accusation from parts of the drone industry: that Skydio was not competing on product quality but rather lobbying to legislate its competitors out of the market. Bry had testified before the House Select Committee on the Chinese Communist Party in June 2024, advocating for restrictions on Chinese drone imports. When the FCC effectively banned new foreign-made drone models from the US market in December 2025, critics were blunt. "Not only is your lobbying crippling existing life-saving technology," one industry expert said, "but also it is hindering innovation at large."

Bry's response was a flat denial. "We had nothing to do with it," he said of the DJI ban legislation. The claim was difficult to square with his Congressional testimony, Skydio's policy apparatus (led by Brendan Groves, a former Associate Deputy Attorney General with deep national security credentials) and the company's relentless positioning as the patriotic alternative. But the underlying national security argument was not frivolous. DJI drones transmit telemetry data through servers that the Chinese government can access. In a military context, that data could reveal troop positions, patrol routes, and operational patterns. The question was whether Skydio's technical superiority in autonomous navigation justified a price premium that most government agencies could not afford, or whether the company was using national security as a shield for a business model that could not survive open competition. The honest answer was that both things were partially true, and that the tension between them would define the American drone industry for years to come.

The ISR Revolution

Skydio's story matters not only because of what the company has built but because of what it represents: the transformation of Intelligence, Surveillance, and Reconnaissance (ISR) from a capability that required expensive, dedicated platforms into something that can be deployed by a single soldier from a backpack. To understand the scale of that transformation, look beyond Skydio to what is happening across NATO.

In January 2025, Sweden unveiled a drone swarm system developed by Saab that enables a single operator to control 100 ISR drones simultaneously. Each drone weighs up to 25 kilograms. The system performs reconnaissance, establishes redundant communication links, and operates in contested electronic warfare environments where GPS and radio signals are degraded. Saab developed the system in 12 months, a timeline that would have been unthinkable in traditional defense procurement, where five years from concept to fielding is considered fast. Personnel can be trained to operate it in days. The system was tested during the Arctic Strike Exercise in March 2025, in the kind of northern European terrain where NATO planners expect a future confrontation with Russia to play out.

In October 2025, Norway went further. The Royal Norwegian Ministry of Defence deployed Valkyrie swarm systems for scalable autonomous ISR, multiple drones operating collaboratively under the control of a single human operator. It was the first operational NATO drone swarm deployment, a milestone that received surprisingly little attention outside defense media. The Norwegian system did what military planners had been theorizing about for a decade: it put genuine multi-drone autonomous reconnaissance into the hands of regular military units, not special operations forces or test programs.

The numbers tell the story of where ISR is headed. In 2025, ISR drones accounted for approximately 40% of all defense drone applications globally. The military drone market was valued at $15.8 billion, projected to reach $22.8 billion by 2030. But the raw market figures understated the operational shift. The traditional ISR architecture, a single MQ-9 Reaper costing $30 million, operated by a crew of two or more, loitering at 25,000 feet, was being supplemented, and in some cases replaced, by swarms of small autonomous drones that cost a fraction as much and could saturate an area with persistent surveillance that a single platform could never match. A Reaper can watch one road. A hundred small drones can watch a city.

The emerging architecture across NATO combines three tiers: high-altitude, long-endurance relay platforms at the top, providing communications and wide-area coverage; mid-altitude tactical drones in the middle, carrying advanced sensors and providing detailed surveillance; and expendable small UAS swarms at the tactical edge, pushed down to platoon level, fused by AI analytics that flag targets and anomalies faster than any human analyst could process the feeds. Skydio's X10D sits at the bottom tier of this architecture, the most numerous, the most expendable, the closest to the fight. It is the drone that a squad leader pulls from a backpack when the patrol needs to know what is around the next corner or over the next hill.

Bry's vision for this tier is what he calls the 1:many ratio. Traditional drones require a one-to-one relationship between operator and aircraft: one skilled pilot per drone. Skydio's autonomy engine is designed to invert that ratio. One operator overseeing five drones. Then ten. Then fifty. "The drone should work for the person, not the other way around," Bry said. "One person should have five, 10, 50 drones doing useful work on their behalf. And the only way you are going to get there is with AI and autonomy, making the thing smart enough to fly itself."

That 1:many architecture is not a future projection. It is already operational. Skydio's Dock system (an automated base station that launches, recovers, and recharges drones without a human physically present) enables persistent, autonomous coverage from fixed positions. Groups of Docks form "hives" that provide 24/7 surveillance with seamless handoff between drones. By February 2026, Skydio's Drone as First Responder platform, which integrates the Dock with autonomous dispatch software, had surpassed 10 million calls for service in the law enforcement market, a number that demonstrates the kind of operational scale that military planners are watching closely.

The Hunter's Edge

The Hunter-Killer framework introduced in this book places every autonomous weapons system on a spectrum between two functions: sensing and striking, finding and killing. Skydio's X10D sits firmly on the "hunter" end of that spectrum. It does not carry weapons. It does not strike targets. It finds them.

But the line between hunter and killer is thinner than it appears. In Ukraine, the tactical pattern is already established and well documented: a Skydio drone or similar ISR platform identifies and locates a target, feeding its coordinates into the Delta battlefield management system. Within minutes, sometimes seconds, a strike drone follows. The hunter does not kill, but without the hunter, the killer is blind. The sensing function and the striking function are becoming two halves of a single automated loop, and the human decision-making that connects them is being compressed into an ever-narrower window.

This is the Compression at work, the shrinking of time between detection and destruction as AI takes over more of the kill chain. And it is advancing faster than most people outside the defense establishment realize. When Bry talks about one operator commanding 50 drones, he means ISR drones. But the same autonomy that lets an ISR drone navigate without GPS, avoid obstacles at 36 miles per hour, and track a target through complex terrain is the same autonomy that a strike drone needs to reach its target in a jammed environment. The technology is dual-use not because anyone designed it that way but because autonomous navigation is autonomous navigation, whether the drone carries a camera or a warhead.

Bry is aware of the tension. "We are on a faster track to artificial general intelligence than much of our civilization realizes," he has said. "But even in a world with AGI, human intent, initiative, empathy, and judgment will remain essential ingredients for ethical conflict." His optimistic view is that future conflicts will have fewer human casualties and become "direct technology and production competitions."

That is one possible future. The other is a world where autonomous sensing and autonomous striking merge into a single platform that compresses the kill chain to zero—where the drone that finds the target is the drone that destroys it, and no human is in the loop at all. The technology to build that world already exists in fragments across a dozen companies and a dozen countries. Whether it gets assembled, and by whom, and under what rules, is the question that the rest of this book will address.

Finding the target is the first problem, and Skydio has built the most capable autonomous solution for it that any Western military has fielded at scale. Destroying the target is the second problem, and in 2026, the weapons designed to close the gap between sensing and striking are proliferating faster than any arms control regime can track. From a testing range outside Ankara to the production lines of the world's most prolific drone exporter, the "killer" half of the spectrum is evolving just as fast.

Chapter 3: The Killers

PART I: THE REVOLUTION

On a testing range south of Ankara, Turkiye, in March 2026, five aircraft lifted off in a staggered sequence that took less than ninety seconds. Each weighed 800 kilograms, heavier than a grand piano, lighter than a compact car. Each carried a 200-kilogram warhead capable of destroying a hardened bunker or a main battle tank. Each had a range exceeding 2,000 kilometers and could loiter for more than 13 hours at 200 kilometers per hour, which meant that a K2 launched from eastern Turkiye could reach targets in the Caucasus, the Persian Gulf, or the eastern Mediterranean and still have fuel to wait.

The Baykar K2 was the largest kamikaze drone in its class. But size and range were not what made this test historic. What made it historic was what happened after the five aircraft reached altitude.

Without a single command from the ground control station, the drones formed an autonomous swarm. Each K2 used its own onboard AI, its own sensors, and its own mission software to calculate and maintain formation position relative to the others. When the swarm detected a simulated target array (three decoy vehicles arranged in a defensive cluster on the desert floor) the five aircraft redistributed autonomously. Two broke formation and descended on a suppression vector, their onboard algorithms identifying the simulated air defense radar as the priority threat.

Three converged from different angles on the primary target. The entire reallocation, from detection to terminal attack geometry, took less than eight seconds. No human touched a control.

Selcuk Bayraktar, Baykar's chief technology officer and the MIT-trained engineer who had led the development of the Bayraktar TB2,the drone that introduced precision strike warfare to a dozen nations that could never have afforded a traditional air force, watched from the control room with his brother Haluk, the company's CEO. The K2 also carried visual terrain-scanning systems that let it navigate in GNSS-denied environments, the kind of electronically contested airspace that had grounded the TB2 in eastern Ukraine before AI upgrades brought it back. The lesson from Ukraine had been absorbed. The next generation would not depend on satellites.

Baykar had exported $2.2 billion in drones the previous year. The TB2 had been sold to 31 countries. The jet-powered Kizilelma unmanned combat aerial vehicle was entering inventory in 2026. And Haluk Bayraktar had announced that the company's keyword for the year was a single word: production.

The K2 represented something new on the Hunter-Killer spectrum. Not a hunter—it carried no sophisticated ISR suite, no persistent surveillance capability. Not a simple killer—it was not a remote-controlled projectile guided by a human thumb on a joystick. It was an autonomous strike system that could find, coordinate, and destroy without human input. And the five-drone swarm test was a proof of concept for formations of dozens, then hundreds. The hunter and the killer were merging inside a single airframe, and the human was stepping out of the loop.

The Bayraktar family's trajectory tells the story of how the killer side of the spectrum evolved. The TB2 completed its maiden flight in August 2014. It was a medium-altitude, long-endurance remotely piloted aircraft, a human watched through its cameras, a human pressed the button. In 2020, TB2s destroyed Armenian armor in Nagorno-Karabakh and Syrian air defenses during Operation Spring Shield, establishing Baykar as the company that gave nations the capabilities of an air force without requiring traditional manned aircraft. Eight Pantsir and Buk air defense systems, Russian-made, multi-million-dollar platforms, fell to TB2 strikes in a single week in Idlib. The drone became a cultural icon in Ukraine, inspiring a folk song after it destroyed Russian command posts, supply trains, and missile systems in the opening weeks of the 2022 invasion. Then Russian electronic warfare adapted, and the TB2 was forced offline for months.

The Bayraktar brothers had built their company from scratch after a US ban on armed drone exports to Turkiye forced Ankara to develop indigenous capabilities. They had created the most produced and export-successful unmanned combat aerial vehicle in the world, expanded production capacity to 500 TB2s per year, opened manufacturing lines in Italy through a joint venture with Leonardo, and established Baykar as the defense company that gave middle powers the tools to fight like great powers. Now, with the K2, they were moving beyond remote control entirely. The arc from the TB2 to the K2 is the arc of the entire killer revolution: from remote-controlled to autonomous, from vulnerable to resilient, from individual platform to coordinated swarm. And Baykar is not alone.

The Killer Ecosystem

The "killer" side of the drone revolution, the systems designed not to watch but to destroy, had become the most crowded and fastest-evolving category in the global weapons market by 2026. To understand how the time between detecting a target and destroying it was compressing toward zero, it is necessary to see the full ecosystem, from the cheapest weapons to the most expensive, from the crudest to the most autonomous.

The ecosystem breaks into four categories, each representing a different point on the spectrum of autonomy, cost, and lethality.

The first category is loitering munitions, weapons that blur the line between a drone and a missile. They launch, fly to a designated area, circle overhead waiting for a target, and then dive to destroy it. The category was invented by Israel in the 1990s with the IAI Harpy, an anti-radiation weapon designed to hunt enemy radar. By 2026, it had become the fastest-growing segment of the global drone market, with systems fielded by more than 40 nations.

AeroVironment's Switchblade family is the benchmark for the American side of this market. The Switchblade 300 is an anti-personnel weapon small enough to fit in a soldier's backpack, tube-launched, and designed to be deployed at the platoon level. The Switchblade 600, its larger sibling, carries an anti-armor warhead derived from the Javelin missile and can destroy a tank. AeroVironment has expanded the family to include a Switchblade 400 variant, meeting demand that surged after Ukraine demonstrated to every military on earth what a small, cheap, expendable weapon could do to a mechanized force. The Switchblade has been fielded by Ukraine, the United States, France, Lithuania, and the United Kingdom.

Anduril Industries offers the Altius family, the ALTIUS-600M for rapid deployment and the ALTIUS-700M for longer endurance and greater payload. Range: up to 273 miles. Speed: 115 miles per hour. Ceiling: 15,000 feet. The Altius can be tube-launched from a C-130 transport aircraft, a UH-60 Blackhawk helicopter, ground vehicles, or larger drones like the MQ-1C Grey Eagle. In September 2023, Anduril tested a live warhead on the Altius-700M and declared it accurate and effective. In August 2025, the first tranche was delivered to Taiwan, a signal of intent that Beijing noticed. The Pentagon awarded a $50 million SBIR Phase III contract for Altius integration with the MQ-9 Reaper, turning the legacy surveillance platform into a loitering munition mothership.

But here is the cautionary tale that the marketing materials leave out. Between 2024 and 2025, Ukraine's Security Service of Ukraine (SBU) abandoned Altius operations entirely. The reason: severe vulnerability to Russian electronic warfare. In the contested electromagnetic environment of eastern Ukraine, where Russian jamming units operate at a density and sophistication unmatched anywhere else on earth, the Altius could not complete its missions. The system that performed flawlessly on a test range in the American desert failed in the one place where performance actually mattered.

The gap between demonstration and combat is the single most important variable in evaluating any weapons system, and the variable that defense marketing is designed to make you forget. A weapon that works in a controlled environment, with cooperative targets and no electronic warfare, proves only that the engineering is sound. A weapon that works in eastern Ukraine, where Russian EW units jam GPS across hundreds of square kilometers, where electromagnetic interference can fry a drone's control link before it crosses the line of contact, and where the target is shooting back, that weapon has proved something else entirely. The Altius proved the engineering. It did not prove the combat capability. And in the killer market, the difference is everything.

No system on the killer side of the spectrum illustrates that gap more vividly than the one Russia built to exploit it.

Russia's Precision Problem

The second category is Russia's answer to the loitering munition challenge: the ZALA Lancet, manufactured by ZALA Aero Group, a subsidiary of the Kalashnikov Group. The Lancet is, by most independent assessments, the most effective Russian precision strike weapon deployed in Ukraine since 2022.

The Lancet features a distinctive double-X wing configuration, an electro-optical targeting system and camera at the nose, and a real-time video link that streams imagery back to the operator's command-and-control station. It is remotely controlled in its terminal phase, the operator watches the target through the drone's camera and guides the weapon to impact.

What makes the Lancet significant is not its sophistication. By Western standards, it is a relatively simple system, no autonomous targeting, no swarm capability, no terrain-following radar. What makes it significant is that it works. Its production scale and effectiveness against high-value targets have forced Ukrainian forces into an escalating arms race of adaptation that reveals more about the killer revolution than any technical specification. Starting in the spring of 2022, Ukrainian troops began building improvised protective cages around their artillery pieces, chain-link fencing, wire mesh, wooden logs, specifically designed to catch and detonate Lancet warheads before they reached the weapon underneath. Imagery from January 2023 showed the rear half of a Lancet tangled in one of these cages, its warhead having failed to detonate. Ukrainian forces deployed inflatable decoys and wooden mock-ups of high-value systems, fake HIMARS launchers, false tank silhouettes, to waste Lancets on targets that cost nothing.

These countermeasures work. Sometimes. But the dynamic they reveal is the one that defines the killer side of the drone revolution: adaptation cycles that compress faster than any previous weapons competition. Ukraine builds cages. Russia adds thermal seekers. Ukraine deploys decoys. Russia upgrades the Lancet's image-recognition algorithms. Ukraine jams the control link. Russia develops fiber-optic guidance. Each cycle takes weeks, not years. Each costs thousands of dollars, not billions. And the advantage oscillates between attacker and defender in a tempo that legacy procurement systems (the kind that take five years to field a new munition) cannot match.

The third category is the one-way attack drone, of which Iran's HESA Shahed-136 is the most prolific example. The Shahed-136 is not a precision weapon. It is a terror weapon and an attrition weapon, a slow, loud, cheap airframe loaded with explosives and launched in salvos designed to overwhelm air defenses through sheer volume. Russia has used Shaheds against both Ukrainian military positions and civilian infrastructure, launching them in waves that force Ukrainian air defense crews to expend interceptors that cost orders of magnitude more than the drone they are shooting down. Iran supplies the airframes; Russia provides the targeting coordinates and the strategic logic. Ukrainian electronic warfare has achieved some success against the Shahed, causing several to fail before reaching their targets, but the weapon's value lies not in its accuracy but in its expendability.

Lose a Shahed, and you have lost an inexpensive airframe. Force the enemy to fire a $500,000 interceptor missile to stop it, and you have won the exchange regardless of whether the drone hits anything. Launch 20 Shaheds in a single salvo, and even if 19 are shot down, the air defense battery has expended $9.5 million in interceptors, and the twentieth gets through. The Shahed inverts the economics of air defense the same way the FPV drone inverts the economics of armored warfare, not through precision but through cost asymmetry. It is the mathematical logic of attrition applied at industrial scale, and it works because the Shahed is cheap enough to lose in bulk and simple enough to manufacture faster than the enemy can produce the weapons to stop it.

The $200 Revolution

The fourth category is the one that changed everything: the first-person view attack drone. The FPV drone is the weapon that put the drone revolution on every screen in the world.

An FPV drone costs approximately $200. It is assembled from commercial racing-drone components, a carbon-fiber frame, brushless motors, a flight controller, a camera transmitter, and a set of first-person-view goggles that let the operator see through the drone's camera in real time. To that commercial platform, a warhead is attached, typically an RPG warhead, a mortar round, a grenade, or an improvised explosive device. The operator flies the drone directly into the target at speeds exceeding 100 kilometers per hour. The drone provides reconnaissance during the entire approach, streaming live video to the operator's goggles until the moment of impact.

$200 drone. $10 million tank. The math is not complicated.

The scale of production is what separates the FPV phenomenon from every previous precision-strike technology. Guided missiles are manufactured in factories with security clearances, multi-year contracts, and unit costs measured in tens of thousands of dollars. FPV drones are assembled on kitchen tables. By October 2023, both Russian and Ukrainian forces were producing thousands of FPV drones every month, many assembled by volunteer groups and civilian workshops operating out of garages, basements, and repurposed commercial spaces. Ukraine produced over 4 million drones in 2024 across all types, aerial, ground, and maritime, with more than 500 producers operating across the country. The target for 2025 was 10 million. Russia, responding to the same logic of mass production, planned 2 million FPV drones and 30,000 long-range drones for 2025. The drone war had become, in the most literal sense, a production competition, and both sides were scaling faster than any defense-industrial base had ever scaled a weapons system before.

The cost asymmetry defines every engagement. In November 2023, Ukrainian soldiers used a civilian-donated FPV drone to destroy a Russian Tor missile system on the Kupiansk front. A Tor costs approximately $24 million. That single engagement, one drone, one operator, one flight, destroyed a system whose price tag equaled 120,000 FPV drones. An air defense system that took years to build, years to integrate into a layered defense network, and months to train crews to operate was burning on the steppe because a volunteer with a soldering iron and a 3D printer had built something that could reach it.

The FPV has also demonstrated reach that its designers never intended. In 2025, during Operation Spider, Ukrainian forces covertly positioned shipping containers within FPV range of Russian bomber airfields deep in Siberia, thousands of kilometers from the front line. The drones launched from the containers, flew short distances to the airfields, and destroyed irreplaceable strategic bombers on the ground. A $200 weapon, forward-deployed in a metal box, struck targets that would normally require cruise missiles or stealth aircraft to reach. The boundary between tactical and strategic had dissolved.

But the FPV drone's most consequential innovation is not its cheapness. It is what happens when cheapness meets tactical coordination. Ukrainian forces developed a standard operating procedure that now defines frontline combat along the entire line of contact: the hunter-killer pair. A reconnaissance drone (often another FPV equipped with a camera rather than a warhead) identifies and locates the target, streaming its position to the strike operator. Within seconds, a strike FPV follows the same path or a different approach vector and executes the attack. Detection to destruction: seconds. The hunter and the killer are separate platforms, but they operate as a single system linked by human coordination and, increasingly, by software.

As of June 2023, Ukraine was losing between 5,000 and 10,000 drones per month, 160 per day, according to Ukrainian soldiers interviewed by defense media. The attrition rate would be catastrophic for any weapon that costs millions. For a weapon that costs $200, it is a production problem, not a strategic one. Build them faster than you lose them, and the math works. That logic, attritable mass, expendable and replaceable at industrial scale, is the principle that every military on earth is now trying to replicate. The country that masters drone production at the scale of consumer electronics, not at the scale of defense procurement, will own the tactical battlefield for the next generation.

The Compression

This is the concept that turns a collection of cheap weapons into a revolution: **the Compression**.

The Compression is the shrinking of time between detecting a target and destroying it. It is the central dynamic of the drone revolution, the force that is squeezing human judgment out of the kill chain.

Consider the historical timeline. In the first Gulf War, the time from a sensor detecting a mobile Scud launcher to a strike aircraft arriving over the target could be measured in hours. Saddam Hussein's forces exploited that gap ruthlessly, moving launchers between detection and destruction. By the second Gulf War, networked command-and-control systems had compressed the detect-to-strike cycle to roughly 45 minutes for time-sensitive targets. By 2010, armed Predator and Reaper drones compressed it further, the same platform that detected the target could strike it, eliminating the handoff between sensor and shooter. But the human was still in the loop, watching the video feed, requesting authorization, waiting for a judge advocate to assess proportionality.

In Ukraine in 2026, the detect-to-strike cycle for an FPV hunter-killer pair is measured in seconds.

The implications of that single fact ripple outward in every direction. The gap that once gave commanders time to deliberate, lawyers time to assess, and targets time to move has been compressed to the point where meaningful human review is becoming operationally impractical at the tactical level. The reconnaissance drone finds the target. The strike drone follows. The operator, typically a 23-year-old in a basement, trained for weeks rather than years: makes the call. There is no judge advocate in that basement. There is no proportionality assessment. There is no after-action review before the next drone is in the air. There is barely time for the operator to confirm that the thermal signature on the screen is a tank and not an ambulance before the drone is committed to its terminal dive.

The Compression also changes the defender's calculus. When the time between being detected and being struck shrinks to seconds, concealment becomes the only survivable strategy. Movement becomes lethal, any vehicle that moves can be spotted, and anything spotted can be struck before it covers 500 meters. The result, visible across the entire Ukraine front, is a paralysis of maneuver. Tanks sit camouflaged in tree lines. Logistics convoys move at night, in small elements, on unpredictable routes. Infantry moves in groups of two or three, not squads. The Compression has turned the open battlefield into a killing ground where exposure is death, and the drone operator is the executioner.

The Compression is not a theory. It is a measurable, observable fact on every kilometer of the Ukraine front line. And the technology to compress it further, to remove the human from the seconds-long gap that remains: already exists.

When the Screen Goes Black

The most consequential adaptation in the Compression is one that most observers outside the defense community have not yet absorbed: automated terminal attack.

In the dense electronic warfare environment of eastern Ukraine, both Russian and Ukrainian jamming systems routinely sever the radio link between an FPV drone and its operator in the final seconds before impact. The screen goes black. The operator's goggles show static. The drone is flying blind, or it was, until engineers on both sides solved the problem by making the drone capable of completing the strike autonomously.

In March 2024, footage surfaced on social media showing a Ukrainian FPV drone being actively jammed in its terminal phase. The video feed degraded, then cut entirely. The operator lost control. The drone struck the target anyway. Its onboard flight controller had been programmed with a terminal-attack algorithm that maintained the drone's trajectory and executed the final dive without human input. The operator set the attack in motion. The machine finished it.

This is the threshold that international arms control lawyers have been debating for a decade, made real not in a DARPA laboratory but in a Ukrainian field workshop with a soldering station and an open-source flight-controller forum. The drone's onboard algorithm completed a lethal strike after the human operator was removed from the loop by electronic warfare. The human initiated the attack. The machine executed it. The target was destroyed by a system operating without human control in its terminal phase.

The legal and ethical questions are not academic. Does the human operator's initial targeting constitute meaningful human control if the machine completes the strike independently? Is the electronic warfare that severs the control link the morally significant act, or is it the engineer who programmed the terminal-attack algorithm? And if automated terminal attack becomes standard (as it increasingly has on both sides of the Ukraine front) does every FPV drone become, in its final seconds, an autonomous weapon? These are the questions that 166 nations voted to address at the United Nations in December 2024. The answers have not arrived. The technology is not waiting.

The second adaptation is fiber-optic guidance. Both Russian and Ukrainian forces have developed FPV drones that trail thin fiber-optic cables, typically 5 to 20 kilometers in length, connecting the drone directly to the operator by physical wire rather than radio signal. The fiber-optic link is completely immune to electronic jamming, no radio frequency to disrupt, no signal to spoof. It does not reveal the operator's position to enemy signals intelligence because it emits no electromagnetic signature. It delivers higher-quality video feeds than radio links, even in heavily contested electromagnetic environments. And because the wire requires less power than a radio transmitter, fiber-optic drones can idle on the ground, powered down, waiting to ambush passing targets with no detectable signature until the moment of launch.

In October 2024, Russian forces deployed FPV drones trailing up to 10 kilometers of fiber-optic wire against Ukrainian positions in Zaporizhzhia Oblast. Ukrainian electronic warfare was useless against them. Ukrainian soldiers resorted to shooting the drones down with small-arms fire, rifles against drones, the oldest counter-air tactic in existence, made necessary because the most advanced electronic countermeasures in the Ukrainian arsenal could not touch a signal traveling through glass.

Fiber-optic drones have limitations. Their range is constrained by the length of the wire. Their payload capacity is reduced by the weight of the spool. They are ineffective in dense urban environments where the cable snags on structures. But their existence demonstrates a principle that applies across the entire killer ecosystem: for every countermeasure, there is a counter-countermeasure, and the adaptation cycle in this war moves at a speed that no traditional procurement process can match.

The Swarm Layer

The final step in the Compression is the one that connects all the individual weapons into something greater than the sum of their parts: autonomous swarm coordination.

Swarmer, the Ukrainian drone-technology company that has become the most combat-tested swarm software provider on earth, has built the coordination layer that turns individual hunter and killer drones into networked systems. Their flagship product, Styx, coordinates reconnaissance and strike drones within autonomous networks, groups of drones that communicate, adapt, and redistribute without direct human control of each individual platform. By September 2025, Swarmer had logged over 82,000 combat missions using Styx. By the time the company filed for its March 2026 IPO on the Nasdaq, the count exceeded 100,000. The stock, listed under the ticker SWMR, surged nearly 1,000% in its first three trading sessions.

Ukraine has tested swarm technologies in more than 100 operations involving groups of 8 to 25 drones. Military analysts say these are the first documented cases of swarm technology used routinely in active combat. The core operational concept for

2026 is moving from "one operator per platform" (the model that still dominates most of the world's militaries) to what Swarmer calls "swarm middleware with intent-based control." The human sets the mission parameters, the rules of engagement, the target priorities. The software handles the coordination. Individual drones are not individually controlled. They are managed.

In Israel, Elbit Systems has developed "Seek & Strike swarms" for the Israel Defense Forces under an approximately $40 million procurement contract. The concept pairs intelligence-collecting drones with attack drones in autonomous formations, what Elbit calls "intelligence platoons." The swarms use advanced ISR to enable adaptive collective behaviors, supported by Elbit's Dominion-X platform, which provides distributed sensing, information fusion, and human-swarm teaming in a single system.

And in Turkiye, Baykar's K2 swarm test was the proof of concept for a vision in which dozens of 800-kilogram autonomous strike platforms coordinate an attack on a defended position without a single human command after launch. Each drone carries a 200-kilogram warhead. Each navigates without satellites. Each redistributes autonomously when the tactical picture changes. The swarm is the weapon. The individual drone is just a component.

The Compression, pushed to its logical conclusion, looks like this: a swarm of autonomous drones detects a target formation, distributes attack roles among its members based on onboard AI assessment of the threat, executes the strike from multiple vectors simultaneously, and assesses the damage, all within seconds, all without human intervention after the initial launch authorization. Find, fix, track, target, engage, assess, the entire kill chain, executed by machines at machine speed.

That capability does not exist as a fielded, integrated system in 2026. But every component of it has been demonstrated separately. Baykar has demonstrated autonomous swarm coordination with five 800-kilogram strike platforms. Swarmer has demonstrated networked hunter-killer operations in more than 100,000 combat missions. Ukrainian engineers have demonstrated automated terminal attack after loss of operator control. Elbit has demonstrated human-swarm teaming with mixed reconnaissance and strike formations operating as autonomous platoons. The individual technologies are proven. The integration challenge (connecting them into a seamless, autonomous kill chain) is an engineering problem, not a conceptual one. And engineering problems, in this domain, are being solved in months rather than decades.

The pieces are on the table. The question is how quickly someone assembles them, and whether any legal, ethical, or strategic framework will be in place when they do.

From Tactical to Strategic

The FPV drones in Ukrainian trenches and the K2 swarms over the Anatolian testing range are rewriting tactical warfare, the close-in, kilometers-from-the-front-line fight where individual weapons engage individual targets. The Compression is real at this scale. It is measurable, accelerating, and already pushing human judgment to the margins.

What the killer ecosystem reveals, taken as a whole, is a transformation that no single system embodies but that all of them together make undeniable. The Switchblade gives a platoon the ability to destroy a tank from a backpack. The Lancet gives a Russian battalion a precision strike capability that does not require calling in air support. The Shahed gives a state sponsor of terror the ability to overwhelm a nation's air defenses through production volume alone. The FPV drone gives a 23-year-old volunteer the destructive reach that once required a multi-billion-dollar air force. And the K2 gives a mid-sized NATO ally the autonomous swarm strike capability that the Pentagon is still testing in laboratories.

Each of these systems compresses the kill chain in a different way. The Switchblade compresses the distance between the infantry squad and the precision strike. The Lancet compresses the time between detecting an artillery piece and destroying it. The FPV compresses the cost of killing a tank from millions of dollars to hundreds. The K2 compresses the role of the human from operator to authorizer, and then, potentially, to spectator. The Compression is not one technology. It is the convergent effect of all of them, accelerating simultaneously, each one pushing the others forward.

But the great powers are thinking at a scale that dwarfs the tactical fight.

On December 11, 2025, at an Aviation Industry Corporation of China (AVIC) test facility, a drone mothership called the Jiu Tian completed its maiden flight. It was 16.35 meters long. Its wingspan stretched 25 meters. Its maximum takeoff weight was 16 metric tons. Its dual internal bays were designed to carry and release approximately 100 small drones, a mix of reconnaissance quadcopters and winged attack munitions. Concept videos released by Chinese state media showed a fleet of Jiu Tians approaching a US Navy carrier strike group in the western

Pacific, opening their bays, and releasing hundreds of autonomous drones to overwhelm the group's defenses through sheer numbers.

The concept video was propaganda. The aircraft was real. And the production goal behind it was one that Beijing had not bothered to hide: one million tactical unmanned aircraft systems by 2026.

The drone revolution is not limited to small, cheap, expendable weapons destroying tanks in Ukraine. It is about to go strategic, projected across oceans, scaled to hundreds of platforms, aimed at the ships and bases that underpin American power in the Pacific. The killers are multiplying, and the motherships are coming to carry them.

Chapter 4: The Motherships

PART I: THE REVOLUTION

On December 11, 2025, at an Aviation Industry Corporation of China (AVIC) test facility in central China, a drone the size of a commuter aircraft lifted off a runway and climbed into gray winter sky. The aircraft had no cockpit. No canopy, no ejection seat, no oxygen system. Where a pilot would have sat, there were bays, two of them, running the length of the fuselage, designed to carry approximately 100 smaller drones packed tight in launch racks. The aircraft was called Jiu Tian. It was 16.35 meters long, with a 25-meter wingspan and a maximum takeoff weight of 16 metric tons. It could carry 6,000 kilograms of payload over a ferry range of 7,000 kilometers and stay airborne for 12 hours.

The maiden flight lasted long enough to prove the airframe, and then the Jiu Tian returned to the runway. The engineers at AVIC recorded the data. Chinese state media recorded the moment. And within weeks, concept videos appeared on platforms controlled or influenced by the People's Liberation Army (PLA) showing what came next: a fleet of Jiu Tian motherships approaching a US Navy carrier strike group in the western Pacific. The motherships open their bays. Hundreds of autonomous drones pour out, reconnaissance quadcopters and winged attack munitions, each executing pre-loaded mission objectives without further human input. The drones fan across the ocean in coordinated waves. They overwhelm the carrier group's defenses through sheer numbers, saturating the radars and missile systems designed to track and destroy individual incoming threats. The escorts fire. They hit some. They do not hit enough.

The videos were propaganda, slickly produced, scored with triumphant music, calibrated for maximum psychological effect on both domestic and foreign audiences. But the aircraft that flew on December 11 was not a rendering. It was metal and composite and fuel and payload capacity. The specifications published after the flight matched or exceeded those of the US Air Force's MQ-9 Reaper in payload and endurance, and the Jiu Tian carried something the Reaper never did: a hundred weapons it could deploy in a single pass.

The scenario the concept videos depicted (autonomous swarms saturating a carrier group's defenses faster than those defenses can respond) is the strategic nightmare driving investment on both sides of the Pacific. The drone revolution that began with $200 FPV weapons in Ukrainian basements had gone strategic. The Hunter-Killer framework that described a soldier guiding a quadcopter into a tank now described a 16-ton mothership delivering 100 autonomous weapons against a $13 billion aircraft carrier. The scale had changed by orders of magnitude. The underlying logic had not changed at all.

Beijing's Million-Drone Bet

The Jiu Tian did not emerge in isolation. It is one component of an industrial-military ecosystem that China has built with a coherence and speed that has no Western equivalent. Understanding that ecosystem requires understanding the model behind it: military-civil fusion.

Military-civil fusion (MCF) is not a slogan. It is an ecosystem design driven from the top, from President Xi Jinping, downward through state-guided universities, research laboratories, and a steadily expanding web of dual-use vendors that can be mobilized for defense priorities while breaking down the contracting barriers that slow Western procurement. Laws compel domestic companies to cooperate with state security organs when requested. The result is a whole-of-nation diffusion model: China takes a capability that begins as a commercial technique, autonomy software, edge computing, computer vision, and routes it through procurement pipelines into military applications, then scales those applications through follow-on contracting, organizational learning, and doctrinal experimentation. A Georgetown Center for Security and Emerging Technology (CSET) study analyzed 2,857 PLA AI-related contract award notices between January 2023 and December 2024 and found a defense industrial base anchored by state-owned enterprises but increasingly complemented by a long tail of nontraditional vendors finding pathways into PLA demand.

In 2020, "intelligentization" (the adoption of AI and advanced technologies across the force) was officially affirmed as a core initiative under Xi's goal of modernizing the PLA by 2035. The drone swarm is a signature application of that strategy.

The China Electronics Technology Group Corporation (CETC) has been developing fixed-wing drone swarms for ground surveillance and precision strike since at least the early 2020s. CETC's Swarm I system debuted at the 2021 Zhuhai air show. Swarm II, unveiled in 2024, upgraded the platform to a top speed of 100 kilometers per hour and more than an hour of flight endurance. The critical technical detail is how these swarms coordinate: rather than relying on a central controller (a single ground station issuing commands to each drone) CETC's systems use onboard algorithms for real-time pathfinding, automated target recognition, and multi-agent reinforcement learning. Each drone processes its own sensor data, classifies its own targets, and coordinates with its neighbors through a distributed protocol. There is no single node whose destruction collapses the swarm.

In 2025, China revealed a 200-drone AI swarm controlled by a single soldier. The system used what Chinese state media described as an "intelligent algorithm" that allowed the individual drones to cooperate autonomously even after losing communication with the operator entirely. The swarm did not stop when the link went down. It continued executing its mission objectives, redistributing tasks among surviving units, adapting to the loss of individual drones without human intervention.

And then there is the production number that stops conversation in Pentagon briefing rooms: Beijing has launched a program to field one million tactical unmanned aircraft systems by 2026. One million. Even if the actual figure falls short (and production targets set by authoritarian governments often do) the ambition reveals the strategic logic. China is not building a drone force. It is building an industrial capacity to produce drones at a scale that makes individual platform losses irrelevant. A million tactical UAS is not an army of expensive, exquisite systems. It is a commodity, expendable, replaceable, attritable in a way that a $2 billion stealth bomber or a $13 billion aircraft carrier is not.

The Jiu Tian mothership is the delivery mechanism for that commodity. Seven thousand kilometers of range means the mothership can launch from mainland Chinese airfields and deliver its payload anywhere in the western Pacific. Twelve hours of endurance means it can loiter, waiting for the right moment or the right target. One hundred drones per sortie, deployed from dual internal bays, means that a squadron of Jiu Tians could put a thousand autonomous weapons in the air in a single wave.

That is the context in which the concept video makes sense. It is not fantasy. It is an engineering roadmap.

The obvious counterargument is that much of what is publicly known about Chinese capabilities comes from air shows and concept videos, demonstrations calibrated for foreign audiences as much as for military utility. Actual operational deployment is harder to verify. China's defense industry has a documented history of overpromising on specifications and underdelivering on reliability. Fair enough. But the counterargument misses the structural point. Even if the Jiu Tian never reaches the specifications AVIC claims, even if the million-drone target slips by years, the institutional architecture is real. The production lines exist. The procurement pipelines exist. The CSET study documented nearly three thousand AI-related military contracts in two years. The question is not whether any single Chinese system works as advertised. The question is whether the ecosystem behind it can generate capability at a pace and scale that the United States cannot match through quality alone.

That is the question that keeps Pentagon planners awake.

The American Response: Exquisite Machines

The United States saw the same trends and reached for a different lever. Where China built mass, America built sophistication. Where Beijing aimed for a million cheap drones, Washington invested in a smaller number of autonomous platforms designed to be smarter, stealthier, and more survivable, and to fly alongside the crewed fighters that remain the backbone of American air power. The program is called Collaborative Combat Aircraft, or CCA, and it represents the most ambitious implementation of manned-unmanned teaming (MUM-T) in history.

The Air Force plans to spend more than $8.9 billion on CCA programs between fiscal years 2025 and 2029. The concept is straightforward in principle and staggeringly complex in execution: build AI-equipped uncrewed combat aircraft, loyal wingmen, that can fly alongside manned fighters, carry additional weapons, perform scouting and electronic warfare, absorb enemy fire, and execute tactical maneuvers without requiring a joystick operator on the ground. The human pilot in the crewed fighter provides intent and oversight. The CCA handles the rest.

Increment 1 of the CCA program moved from concept to flight hardware faster than almost any modern defense program. Two companies won the competition. General Atomics, the San Diego-based builder of the Predator and Reaper families, developed the YFQ-42A. Anduril Industries, Palmer Luckey's defense-tech company, valued at $60 billion by March 2026, developed the YFQ-44A, designated "Fury." The Fury's lineage traces to Blue Force Technologies, a small North Carolina firm that Anduril acquired specifically for its stealthy airframe design. General Atomics brought decades of experience building large uncrewed aircraft. Anduril brought the software-first approach of Silicon Valley and a 17-foot-wingspan stealth drone already in flight testing. The YFQ-42A completed its first flight in August

2025. The Fury followed on October 31, 2025, Halloween, as Anduril's engineers noted with the dark humor endemic to defense work. From contract award to first flight in roughly 18 months, a timeline that would have been considered impossible under traditional Pentagon procurement.

The Air Force expects to procure between 100 and 150 Increment 1 CCAs. The broader program could eventually reach the low thousands of airframes. The target cost per unit is between $25 million and $30 million, roughly one-quarter to one-half the cost of an F-35 Lightning II. That ratio is the economic logic of the entire program: field enough autonomous wingmen that losing some in combat is acceptable, at a cost that does not bankrupt the force structure. The defense term for this is **attritable**: designed to be lost. A $25 million CCA is not cheap. But it is cheap enough to risk in a way that a $180 million F-22 Raptor is not, and it does not carry a pilot whose loss creates a training replacement cycle measured in years.

The most telling indicator of urgency is where the CCA control systems are being installed first. The F-22 Raptor,America's premier air superiority fighter, a platform so advanced that Congress prohibited its export to any foreign nation, is being upgraded with new tablet-based control systems that will allow its pilots to direct CCA wingmen in flight. Installation begins in fiscal year 2026. The F-22 will be the first operational platform to control CCAs. The stealth fighter designed to dominate the sky is being retrofitted to command robots.

The Sixth-Generation Bet

The CCA program does not exist in isolation. It is one element of a larger architecture whose centerpiece is the most expensive fighter aircraft program in American history: the F-47.

On March 21, 2025, the Department of the Air Force announced that Boeing had won the engineering and manufacturing development contract for the Next Generation Air Dominance (NGAD) platform, the first US sixth-generation fighter, designated the F-47. The contract was valued at more than $20 billion. The Air Force intends to procure at least 185 aircraft. The F-47 will have a combat radius exceeding 1,000 nautical miles, a top speed above Mach 2, and (according to Air Force Chief of Staff General David Allvin) "significantly longer range, more advanced stealth, be more sustainable, supportable, and have higher availability than our fifth-generation fighters." Manufacturing of the first F-47 began in September 2025. First flight is expected in 2028, with the aircraft becoming operational by 2029.

The F-47 exists because of a conclusion that Air Force leaders reached after a sweeping internal study in 2024, prompted by cost overruns that pushed the per-unit price to roughly three times that of an F-35. The study evaluated alternative force structures, more F-35s, uncrewed-only options, distributed approaches. The result was unambiguous. "Bluntly, what this study told us, we tried a whole bunch of different options, and there was no more viable option than NGAD to achieve air superiority in this highly contested environment," said Major General Joseph Kunkel, director of Air Force Force Design, Integration, and Wargaming. General Kenneth Wilsbach of Air Combat Command was more direct: the United States would need crewed sixth-generation aircraft to counter Chinese sixth-generation aircraft.

The F-47 is designed from inception to operate as the hub of a manned-unmanned team. The original Penetrating Counter-Air concept envisioned a solo crewed fighter. But as autonomous technology advanced, the architecture shifted. In its current design, the F-47 will operate with up to five CCA wingmen, a single human pilot commanding a formation of six aircraft, five of them autonomous. The math of that formation is significant: 185 F-47s, each commanding five CCAs, yields a force of more than 1,100 aircraft, the majority of them expendable. Add the F-35 fleet, which will also integrate CCA support, and the B-21 Raider stealth bomber and E-7 airborne command aircraft, and the picture that emerges is an air force where crewed platforms are the minority and autonomous systems provide the mass.

That is a fundamental inversion of how American air power has worked since the Korean War. For seven decades, the defining constraint of American combat aviation has been the pilot: hard to train, impossible to replace quickly, physiologically limited to nine G-forces and psychologically limited by the processing speed of a human brain under combat stress. Remove the pilot from most of the formation and the force structure changes in ways that extend beyond simple numbers. Autonomous wingmen do not experience fatigue. They do not require years of flight school. They do not create Gold Star families when they are shot down. The CCA is not just a cost savings measure. It is a redefinition of what constitutes an air force.

The question is whether America can build these systems fast enough. Increment 2 of the CCA program is already in planning, with the Air Force's Futures command leaning toward a "low end" CCA, prioritizing affordability and mass over exquisite capability. The lesson of Ukraine, absorbed slowly but now clearly, is that quantity has a quality all its own, and that a force composed entirely of expensive, irreplaceable platforms is a force that can be attrited into irrelevance by an adversary willing to spend cheap.

Every Service, Every Domain

The Air Force is not alone. Every branch of the US military is building its own version of the autonomous wingman concept, and the speed of that expansion reveals how deeply the drone revolution has penetrated American defense planning.

On January 8, 2026, the US Marine Corps awarded a $231.5 million contract to Northrop Grumman and Kratos to develop its first operational CCA under the Marine Air-Ground Task Force Uncrewed Expeditionary Tactical Aircraft program. The contract transitions the Kratos XQ-58 Valkyrie from an experimental testbed into an operational loyal wingman for Marine F-35s.

The XQ-58 Valkyrie is a different kind of drone from the small FPV weapons of Ukraine. It is a stealthy, runway-independent combat aircraft with a maximum takeoff weight of 6,000 pounds, a service ceiling of 45,000 feet, and a range of approximately 3,000 nautical miles. It does not need a runway. It launches from a rocket booster mounted on a truck or a shipping container, the kind of asset that can be hidden, dispersed, and operated from austere locations that a conventional fighter base cannot. The Marines are developing the MQ-58B electronic attack variant specifically for the Suppression of Enemy Air Defenses (SEAD) mission, the extraordinarily dangerous job of destroying an adversary's radar and missile systems so that other aircraft can operate. It is the kind of mission where losing the aircraft is likely and losing the pilot is unacceptable. An attritable autonomous platform changes that calculus entirely.

The Navy's contribution arrived a month earlier. In December 2025, at the Point Mugu Sea Range off the California coast, the US Navy demonstrated manned-unmanned teaming for its own CCA initiative. Two BQM-177A subsonic aerial targets, fast, maneuverable jet drones normally used as training targets for ship-based air defense systems, were flown autonomously using Shield AI's Hivemind software. Hivemind is the artificial intelligence piloting system developed by Brandon Tseng's company, designed from inception for GPS-denied, communications-denied, human-input-denied operations. The demonstration proved that an AI could autonomously fly combat-representative aircraft in a naval environment, a capability the Navy intends to operationalize through additional fleet exercises planned for 2026.

Tseng's path to that demonstration traced back a decade, to a garage in San Diego where he and his brother Ryan debugged the first prototype of a quadcopter called Nova, designed to fly autonomously into buildings and map hostile interiors without GPS or a human pilot. Tseng conceived the company after a mission in Afghanistan's Uruzgan province where his SEAL unit suffered casualties because they lacked adequate reconnaissance of a hostile building. The gap between what a commander needed to know and what technology could tell him, that was the problem Hivemind was built to solve. By December 2025, the AI software born from that garage was autonomously flying Navy combat aircraft at a sea range off the California coast.

The convergence is unmistakable. The Air Force is building CCA wingmen for its F-22s and F-47s. The Marines are building rocket-launched autonomous attack drones for their F-35s. The Navy is proving that AI can fly combat aircraft without a human at the controls. Three services, three different programs, three different contractors, and a single underlying thesis: the future of American military aviation includes platforms that fly, fight, and, if necessary, die without a pilot aboard.

The Carrier-Killer Scenario

Now put the two halves together.

On one side: China's Jiu Tian motherships, each carrying 100 autonomous drones, supported by a million-unit production base, coordinated by CETC's multi-agent reinforcement learning algorithms, operated under a military-civil fusion model that compresses the path from laboratory to battlefield. On the other: America's carrier strike groups, the most powerful naval formations in human history, centered on $13 billion nuclear-powered aircraft carriers carrying 70-plus manned aircraft and escorted by guided-missile cruisers and destroyers armed with the Aegis combat system.

The scenario that drives investment on both sides of the Pacific goes like this. A squadron of Jiu Tians launches from airfields on the Chinese mainland. They fly 2,000 kilometers east into the Philippine Sea, beyond the range of carrier-based fighter patrols. At a pre-programmed release point, the motherships open their bays. Five hundred autonomous drones—maybe more—disperse into coordinated waves. Some carry reconnaissance sensors, mapping the carrier group's defensive posture in real time. Others carry electronic warfare payloads, jamming the escorts' radar and communication links. The attack drones, the killers, follow the hunters. They approach from multiple vectors simultaneously, at low altitude, presenting hundreds of small radar cross-sections instead of a few large ones. The Aegis system tracks and engages. It destroys dozens. It does not destroy hundreds. The math that turned a $200 FPV drone into a tank killer operates at oceanic scale: enough cheap weapons, arriving fast enough, from enough directions, will overwhelm any defense designed to engage threats one at a time.

This scenario is not inevitable. Responsible analysts do not treat it as a prediction. It is a planning construct, the strategic context that explains why the US Navy is investing in autonomous wingmen, why the Air Force is spending $8.9 billion on CCAs, why the Marines want rocket-launched drones that do not require runways, and why the F-47 exists at all. The carrier strike group remains the most versatile and powerful naval formation ever assembled. Its defenses are layered, redundant, and continuously upgraded. The Aegis system, the Standard Missile family, the SeaRAM close-in weapon system, and electronic warfare suites represent decades of investment in exactly the problem of defending against saturation attacks.

But the economics are moving against the defense. A single Standard Missile-6 costs approximately $4.3 million. A single autonomous attack drone from a Jiu Tian bay might cost a few thousand dollars, perhaps tens of thousands for a more sophisticated variant. When the cost of the interceptor exceeds the cost of the incoming threat by a factor of a hundred or more, the mathematics of attrition favor the attacker. This does not mean carriers are obsolete. It means their operating environment is changing in ways that demand new defensive concepts, new force structures, and new doctrine, all of which are being developed, in parallel, on both sides of the Pacific.

There is a further complication: the mothership itself is a vulnerability. Destroy the Jiu Tian before it opens its bays, and the hundred drones inside die with it. A 16-ton, subsonic, uncrewed aircraft is not invisible to a carrier group's sensors. American air superiority fighters, F-35s today, F-47s tomorrow, could intercept motherships before they reach their release points. This is the counterargument that defense optimists make, and it has merit. Distributed launch, from trucks, shipping containers, surface vessels, submarines, may prove more survivable than centralized carrier platforms. The drone swarm that launches from a hundred different points is harder to defeat than one that launches from five large motherships.

But centralized and distributed are not mutually exclusive. The Jiu Tian may prove most valuable not as a delivery system for naval attack but as a long-range force projection platform for land campaigns, or as a decoy that forces the adversary to expend expensive interceptors while distributed swarms attack from other vectors. The strategic significance of the mothership concept is not the specific airframe. It is the idea—that drone swarms can be projected across oceanic distances, that the hunter-killer architecture scales from meters to thousands of kilometers.

The honest assessment is this: nobody knows whether the carrier-killer scenario would work as depicted in Chinese concept videos. Real warfare is messier than simulations. Electronic warfare cuts both ways. Autonomous systems fail in ways their designers did not anticipate. The uncertainty and chaos of combat apply to algorithms as surely as they apply to admirals. What the scenario does reveal is the strategic logic driving both nations' investment, and the speed at which that logic is converting from theory into hardware.

The Scale of the Spectrum

Step back and look at what the first four chapters of this book have described. In the basements of Zaporizhzhia, Ukrainian operators guide $200 FPV drones into Russian tanks. At Skydio's factory in Hayward, California, autonomous reconnaissance drones come off the line at more than a thousand per month. At testing ranges in Turkiye, five-drone swarms execute coordinated strikes without human commands. At AVIC facilities in China, a 16-ton mothership carries 100 autonomous weapons 7,000 kilometers. At Boeing's plant in St. Louis, engineers are building a $20 billion sixth-generation fighter designed to command five AI wingmen. At Point Mugu, off the California coast, an AI called Hivemind flies Navy combat drones without a human at the controls.

The Hunter-Killer framework applies at every point on that spectrum. A $200 FPV and a 16-ton Jiu Tian operate on the same logic: find the target, close the gap, execute the strike. The difference is scale, tactical versus strategic, hundreds of meters versus thousands of kilometers, a single tank versus a carrier strike group. And at every scale, the same question applies: where does the human lose meaningful control? At what point in the chain from detection to destruction does the speed of autonomous execution exceed the speed of human judgment? The $200 FPV operator still guides the drone in its final seconds. The CETC swarm keeps operating after the operator's link goes dead. The Jiu Tian concept video shows drones executing mission objectives without further human input. The Compression, the shrinking gap between hunter and killer: is not a single event. It is a gradient, and every system described above sits somewhere on it.

The revolution is real. From Ukrainian basements to Chinese test facilities to American defense laboratories, the evidence is consistent and overwhelming. Autonomous weapons are being designed, tested, funded, and fielded at a pace that has no precedent in the history of military technology.

But understanding that the revolution is happening is not the same as understanding how it works. How does an AI system navigate without GPS? How do 50 drones coordinate without a central controller? How does a machine distinguish a tank from a tractor, a combatant from a civilian? The answers to those questions determine not just what these weapons can do, but what limits, if any, can be placed on what they will do. Part II takes the reader inside the technology that makes autonomous warfare possible, beginning with the artificial intelligence that serves as the brain of every system described so far, and reveals why understanding it is the prerequisite for governing it.

Chapter 5: The Brain

PART II: THE TECHNOLOGY

San Diego, California, 2015. Brandon Tseng sat on a folding chair in a garage, staring at a quadcopter that would not fly straight. The drone, a prototype called Nova, assembled from commercial components and custom software running on a processor the size of a credit card, kept drifting left, overcorrecting, then drifting left again. Tseng's brother Ryan knelt on the concrete floor, laptop balanced on a plastic storage bin, rewriting navigation code. Their co-founder Andrew Reiter was soldering a replacement motor mount at a workbench made from a door laid across two sawhorses. They had $100,000 in seed money gathered from friends and family. They had no office, no employees, and no contract. What they had was a problem.

Two years earlier, Tseng had been a US Navy SEAL deployed to Afghanistan's Uruzgan province. During a mission to clear a compound, his unit had taken casualties because they lacked adequate reconnaissance of a hostile building. They went in without knowing what was inside. The gap between what the commander needed to know and what available technology could tell him, that gap had cost lives. Tseng came home with a Bronze Star, a Combat Action Ribbon, and a conviction that the problem was solvable. An aircraft small enough to fly through a doorway, smart enough to navigate a building without GPS or a human pilot, and fast enough to map hostile interiors in real time, that was what his teammates had needed. That was what he intended to build.

The garage in San Diego was where it started. The drone they were debugging was designed to do something no military system could do in 2015: fly autonomously into an unknown building, navigate rooms and hallways without satellite signals or a radio link to an operator, build a map of the interior in real time, and report what it found. No GPS. No communications tether. No human at the controls. The Nova had to figure out where it was, where it was going, and what it was looking at, all by itself, all on a processor that drew less power than a desk lamp.

A decade later, the artificial intelligence software they developed in that garage, Hivemind, would autonomously fly US Navy combat aircraft in a manned-unmanned teaming demonstration off the coast of California. Shield AI, the company that grew from $100,000 in seed funding to a $5.3 billion valuation in March 2025, reaching $5.6 billion by late 2025, would have more than 1,000 employees and contracts with the US Air Force, Navy, Marine Corps, and Special Operations Command. The Nova would become the first AI-powered drone deployed for defense purposes in US military history. And the problem Tseng set out to solve in that garage (how to build a machine that operates when every link to the outside world is severed) would become the defining technical challenge of autonomous warfare.

That challenge has a name. It is called edge AI. And understanding it is the prerequisite for understanding everything that follows.

The Intelligence at the Edge

Edge AI is a simple concept with enormous consequences. It means the artificial intelligence runs on the drone itself, not in a cloud server, not relayed through a satellite link, not processed in a ground station and transmitted back to the aircraft. The AI lives on the hardware it controls, processing sensor data and executing decisions in real time, right there at what engineers call the edge of the network.

This matters because the modern battlefield is designed to sever connections. Russian electronic warfare units demonstrated the cost of dependence on remote links early in the Ukraine conflict. A study by the Royal United Services Institute (RUSI) found that Russian EW destroyed or disabled approximately 90% of Ukrainian drones during the spring of 2022, not by shooting them down but by jamming their GPS signals and severing their radio links to operators. A drone that relies on a satellite connection for navigation is a drone that can be blinded. A drone that requires a radio link for targeting is a drone that can be deafened. A drone that needs a human operator to issue real-time commands is a drone that can be disconnected from its commander and left to crash or wander.

Edge AI eliminates that vulnerability. When the intelligence runs on the drone's own processor, jamming the radio link does not stop the system from navigating. Spoofing the GPS signal does not prevent it from knowing where it is. Cutting communications does not prevent it from completing its mission. The drone becomes, in a meaningful engineering sense, self-sufficient. Not intelligent in the way a human is intelligent, it does not understand, reason, or intend. But operationally autonomous: capable of sensing its environment, processing what it senses, and acting on that processing without waiting for instructions from anywhere else.

The hardware that makes this possible has undergone a revolution of its own. Modern autonomous drones carry single-board computers (processors no larger than a smartphone, drawing a few watts of power) that run machine learning models trained on millions of images. The NVIDIA Jetson Orin, the processor inside the Skydio X10D, runs nine custom deep neural networks simultaneously, building a three-dimensional obstacle map of the drone's surroundings at more than one million points per second. That is not remote control with stabilization software. That is a machine constructing a model of the world around it and navigating through that model in real time. Every drone described in Part I of this book, from the simplest Ukrainian FPV with autonomous terminal guidance to the Jiu Tian mothership deploying 100 autonomous weapons, depends on some version of this capability. The processor at the edge is what makes the difference between a remote-controlled aircraft and an autonomous one.

Shield AI's Hivemind was built from the ground up for the hardest version of this problem. Its design specification reads like a list of everything a battlefield can take away: "No GNSS, no comms, no human input required." The software does not treat GPS denial or communications loss as failure conditions to recover from. It treats them as the baseline operating environment. Hivemind uses onboard sensors—cameras, inertial measurement units, and proprietary algorithms—to navigate, avoid obstacles, identify objects of interest, and coordinate with other Hivemind-equipped aircraft, all without any external signal. When Brandon Tseng's team debugged that first Nova prototype in a San Diego garage, they were not building a drone that could fly in a nice-to-have autonomous mode. They were building a drone that assumed the operator's link would be severed, the satellites would be jammed, and the human would be unavailable—and designed every line of code for that reality.

That assumption is what separates edge AI from automation. An automated drone follows pre-programmed waypoints. It knows the route before takeoff and executes it. An edge AI system perceives its environment in real time, adapts to changes it did not anticipate, and makes decisions its programmers did not specifically pre-plan. The distinction is between a train on tracks and a car that drives itself through traffic. Both move without a human pushing them, but only one responds to the world as it finds it.

How a Drone Sees

The technical foundation of edge AI is a cluster of capabilities that, taken together, give a machine something analogous to spatial awareness. They are not consciousness. They are not understanding. They are pattern recognition and mathematical optimization operating at extraordinary speed. But they allow a drone to do what Brandon Tseng's teammates needed: enter an unknown space and make sense of it without human help.

The first capability is Simultaneous Localization and Mapping, or SLAM. SLAM is a computational method that allows a machine to construct a map of an unknown environment while simultaneously tracking its own position within that map. It is the same class of technology that makes self-driving cars work, adapted for three-dimensional aerial navigation. A SLAM-equipped drone uses its cameras to identify visual features in the environment (a doorframe, a window, the corner of a building, the edge of a rooftop) and triangulates its own position relative to those features. As the drone moves, the map grows. As the map grows, the drone's understanding of its position becomes more precise. The system bootstraps itself: the map helps it navigate, and navigation helps it build the map.

Skydio demonstrated this capability in one of the most vivid proofs of concept in the drone industry. Engineers flew an X10D through a parking garage, a GPS-denied environment with concrete pillars, low ceilings, and no satellite signals of any kind. The drone navigated the structure autonomously, building a 3D map in real time, avoiding obstacles it had never seen before, and returning to its starting point without ever receiving a GPS fix or a command from a human operator. The six navigation cameras arranged around the X10D's body, three on top, three on the bottom, providing 360-degree visual coverage, fed data into the Autonomy Engine at rates that allowed the drone to react to a new obstacle in less time than a human blink.

The second capability is visual-inertial odometry, or VIO. Where SLAM builds a map, VIO tracks the drone's own movement through that map by fusing visual data from cameras with acceleration and rotation data from an inertial measurement unit. The combination is powerful: the camera tells the drone what it sees, the IMU tells the drone how it is moving, and the fusion of both gives the system a continuous, high-frequency estimate of its position and orientation. On the Skydio X10D, VIO operates reliably at altitudes up to 300 meters and enables flight in environments that would defeat any GPS-dependent system, bridge undersides, warehouse interiors, industrial facilities, and contested battlefields where electronic warfare has rendered satellite navigation useless.

The third capability is computer vision, the ability of a neural network to identify and classify objects in a camera feed. This is where edge AI becomes not just navigationally autonomous but tactically relevant. A drone that can navigate without GPS is useful. A drone that can navigate without GPS and distinguish a T-72 tank from a farm tractor is a weapon. Computer vision systems trained on millions of labeled images can classify objects at speeds and accuracy levels that exceed human performance under certain conditions, particularly conditions involving poor visibility, rapid motion, or the need to scan large areas quickly. The limitation is equally important: these systems classify based on visual patterns they have been trained to recognize. They do not understand what a tank is. They match pixel patterns against a database. When the target matches the training data, the system performs well. When the target does not match, when it is obscured, camouflaged, or simply different from anything the system has seen before, performance degrades in ways that are difficult to predict and impossible for the system itself to recognize.

This triad (SLAM for mapping, VIO for positioning, computer vision for identification) is the perceptual architecture of edge AI. It is what allows a Skydio X10D to fly autonomously through a contested urban environment. It is what allows a Shield AI Nova to map the interior of a hostile building. And it is what allows the next generation of autonomous weapons to operate when every connection to the outside world has been cut.

When the AI Fights Back

In August 2020, inside a simulation environment at a Defense Advanced Research Projects Agency (DARPA) facility, an artificial intelligence agent built by a small Virginia company called Heron Systems fought an experienced human F-16 pilot in simulated within-visual-range air combat. The engagement was part of DARPA's AlphaDogfight Trials, a competition designed to test whether AI could match a trained fighter pilot in the most demanding form of aerial warfare: one-on-one dogfighting, close range, no missiles, guns only.

The AI won. Five engagements, five victories. 5-0.

The human pilot, a graduate of the Air Force's weapons instructor course, the military's top-tier tactical aviation program, was outmaneuvered in every engagement. The AI agent executed instantaneous calculations of closure rate, angle of attack, and energy state that no human brain could match. It did not experience G-force-induced tunnel vision. It did not hesitate. It did not second-guess a maneuver that felt risky. It processed the geometry of the fight and executed the optimal response, every time, at machine speed.

The AlphaDogfight Trials were a simulation. The next step was not. Beginning in December 2022, DARPA's Air Combat Evolution (ACE) program moved from the simulator to the sky. The vehicle was the X-62A VISTA, a heavily modified F-16 capable of autonomous flight, operated by the Air Force Test Pilot School at Edwards Air Force Base, California. Over the next nine months, the ACE team flew 21 test flights, progressively building from basic defensive maneuvers to aggressive offensive engagements to the scenario that military aviation planners had been discussing for years but never seen in practice: an AI-controlled fighter jet autonomously dogfighting a human-piloted F-16 in real airspace, at real speeds, with real consequences for failure.

The flights culminated in nose-to-nose passes at closing speeds of 1,200 miles per hour, with the two aircraft separated by as little as 2,000 feet. At those speeds and distances, the margin for error is measured in fractions of a second. A miscalculation by either aircraft, the AI or the human, could result in a midair collision. The AI performed. It maneuvered aggressively, tracked the human fighter's position, and executed tactical decisions that required the kind of split-second judgment that military aviation has spent decades training humans to develop.

The ACE program did not prove that AI is ready to replace human fighter pilots. The engagements were controlled, the scenarios were constrained, and the safety systems were extensive. What ACE proved was something more foundational: that AI can operate a high-performance combat aircraft in dynamic, adversarial conditions at speeds that exceed human cognitive limits. The reaction time required to process a closing speed of 1,200 miles per hour and execute a tactical response is not a reaction time the human nervous system can consistently match. The AI can.

That finding, demonstrated, not theoretical, laid the groundwork for the Collaborative Combat Aircraft program described in Chapter 4. The CCA does not require AI to replace the human pilot. It requires AI to fly alongside the human pilot, executing maneuvers and making tactical decisions in the portion of the engagement envelope where human cognition is too slow. The F-22 Raptor being retrofitted to command CCA wingmen, the F-47 designed to control five autonomous aircraft, the Marine Corps XQ-58 Valkyrie, every one of these programs traces its technological lineage to the moment when a Heron Systems algorithm beat an experienced fighter pilot 5-0 in the AlphaDogfight Trials and then proved it could do the same thing in a real airplane.

The Spectrum of Autonomy

Shield AI's Hivemind, Skydio's Autonomy Engine, and DARPA's ACE represent different points on a spectrum that Paul Scharre first mapped in his 2018 book *Army of None*. Scharre, a former Army Ranger and current executive vice president at the Center for a New American Security (CNAS), established the vocabulary that the defense community still uses to describe machine autonomy. His framework identified three levels of human involvement: human in the loop (a person authorizes every action), human on the loop (a person monitors and can intervene), and human out of the loop (the system acts without human involvement).

That framework remains essential. But the technology has moved faster than the vocabulary. By 2026, the binary distinctions Scharre drew have become a gradient, and the gradient is more compressed, more populated, and more consequential than anyone anticipated in 2018. The autonomy spectrum, updated for the systems that exist now, looks like this.

At one end: remote-controlled. One human operator, one drone, full manual control. The operator sees through the drone's camera, moves the drone with a joystick, and decides when and where to strike. This is the standard operating model for first-person view (FPV) combat drones in Ukraine, the $200 weapons that have dominated the front lines since 2023. The human is in the loop at every moment. The drone does nothing the operator does not command.

One step further: automated. The drone flies a pre-programmed route along GPS waypoints. The human sets the mission before launch and monitors from a ground station. If the drone encounters something unexpected, it follows pre-set rules, return to base, hold position, continue along the route. This is the operating model for most legacy Intelligence, Surveillance, and Reconnaissance (ISR) platforms, including earlier-generation systems like the MQ-1 Predator. The human is on the loop, supervising, not commanding every movement.

Further still: semi-autonomous. AI handles navigation, obstacle avoidance, and target identification in real time. The human approves or denies specific actions: particularly weapons release. The system operates independently for everything except the most consequential decisions. This is where the Skydio X10D lives for ISR operations: the drone navigates autonomously through GPS-denied environments, identifies objects of interest, and presents them to an operator for decision. The Anduril Fury CCA prototype occupies this space as well, designed to fly alongside a crewed fighter, executing tactical maneuvers autonomously, but with a human pilot providing intent and authorizing engagement. The human is still on the loop, but the loop is getting faster, and the window for human review is shrinking.

And at the far end: fully autonomous. The system acts independently using onboard AI and machine learning. No human input required during the mission. Shield AI's Hivemind is designed for this end of the spectrum, explicitly built for conditions where GPS is denied, communications are severed, and no human is available to provide guidance. The Baykar K2 kamikaze drone, unveiled in March 2026, uses its own AI, sensors, and software to maintain swarm formation and execute mission objectives without human input. The Fortem DroneHunter 5.0, an autonomous counter-drone interceptor that began deliveries in January 2026, coordinates up to five interceptors against five simultaneous threats with zero human involvement through its SkyDome command system.

The critical observation is not that fully autonomous systems exist, defensive systems like the Phalanx close-in weapon system and Israel's Iron Dome have operated autonomously in engagement for decades. The critical observation is the speed at which the spectrum is compressing. Systems that were semi-autonomous two years ago are pushing toward full autonomy. Systems that required human approval for weapons release are being tested in modes where the engagement window is too short for a human to meaningfully review the decision. The space between "human on the loop" and "human out of the loop" is measured in milliseconds, and the gap is narrowing with every software update, every training iteration, every combat deployment that demonstrates the cost of waiting for a human to approve.

What Happens When the Link Goes Dead

The question that haunts every autonomous weapons engineer is not what happens when the system works as designed. It is what happens when the environment takes away the things the system was designed to have.

In Ukraine, that question has been answered, repeatedly, violently, and in real time. Russian electronic warfare is the most sophisticated and densely deployed EW environment in the history of armed conflict. The jamming is not a single frequency or a single technique. It is a layered architecture of GPS spoofers, radio-frequency jammers, and directional energy systems designed to sever every link between a drone and its operator. Early in the war, that architecture was devastatingly effective. Ukrainian drone units reported losses of 90% or more to EW before the drones could reach their targets.

The Ukrainian response drove the development of edge AI capabilities faster than any Pentagon program could have managed. When GPS jamming rendered satellite-dependent navigation useless, Ukrainian engineers added visual navigation, cameras feeding SLAM algorithms that allowed drones to navigate by matching terrain features against pre-loaded maps. When radio-frequency jamming severed the operator's control link during the final approach to a target, Ukrainian developers implemented autonomous terminal guidance, AI that takes control of the drone in its last seconds, using computer vision to identify and track the target, completing the strike even after the human operator's screen goes black. In March 2024, combat footage showed a Ukrainian FPV drone continuing its attack run after being actively jammed, striking its target without any human input during the final phase of flight. The operator had already lost the signal. The AI finished the job.

Fiber-optic drones represent another response to the same problem. By trailing a thin optical cable behind the drone, cables running 5 to 20 kilometers in length, operators maintain a communication link that is physically immune to radio-frequency jamming. The signal travels through glass, not air. Russian electronic warfare cannot jam a fiber-optic cable. These drones sacrifice some range and maneuverability for an unjammable link, and they can idle on the ground in ambush positions, drawing no electronic signature until activated. Both Russia and Ukraine now deploy fiber-optic FPV drones in significant numbers.

But the more consequential response is what happens when all links fail, when GPS is spoofed, communications are jammed, fiber-optic cables are not available, and the drone is entirely on its own. This is the scenario that Hivemind was built for, and it is the scenario that defines the cutting edge of autonomous weapons development in 2026.

The answer involves a hierarchy of autonomous fallback behaviors. First: pre-mission intent loading. Before launch, the drone receives its mission objectives, target coordinates, flight path, rules of engagement, and contingency instructions. If communications are lost, the drone executes the pre-loaded mission autonomously. It does not wait for new orders. It does what it was told to do before the link went down.

Second: decentralized consensus. In swarm operations, if the base-station link is severed, the drones continue operating by communicating with each other through inter-drone mesh links. The swarm does not need a central controller. Each drone shares its sensor data with its neighbors, and the group reaches agreement on task allocation, threat response, and mission execution through consensus algorithms that distribute decision-making across the entire formation. When the Chinese CETC swarm demonstrated 200 drones controlled by a single soldier in 2025, the critical capability was not the control link to the operator, it was the system's ability to continue executing after the operator's link was cut. The swarm kept operating, redistributing tasks among surviving units, adapting to the loss of individual drones without human intervention.

Third: autonomous fallback behaviors. If all communications are lost, no base link, no inter-drone mesh, total isolation, the drone reverts to pre-programmed behaviors. Return to base. Orbit at a designated point. Continue executing the last mission segment. The specific behavior depends on the mission profile and the rules loaded before launch. The drone does not improvise. It follows the last set of instructions it received.

The Baykar K2 kamikaze drone, Turkiye's largest autonomous strike platform unveiled in March 2026, demonstrates the full hierarchy. With a maximum takeoff weight of 800 kilograms, a 200-kilogram warhead, a range exceeding 2,000 kilometers, and more than 13 hours of endurance, the K2 can estimate its position through visual terrain scanning using a gimbal camera and night-vision lower camera, advancing autonomously toward its target even in the total absence of GPS signals. In swarm formation tests with five aircraft, each platform used its own AI, sensors, and software to maintain position and execute objectives without human input.

This is the practical reality of the Compression. The gap between hunter and killer narrows not because engineers want to remove humans from the decision loop, but because the battlefield removes them. Electronic warfare severs the link. Jamming closes the window for human review. The speed of the engagement exceeds the speed of human cognition. The AI does not seize control. The environment forces the AI to take it.

The Reliability Problem

The strongest counterargument against edge AI in autonomous weapons is not ethical: it is practical. The systems are not reliable enough.

Dario Amodei, CEO of Anthropic, one of the world's leading AI companies, stated the objection directly: "Frontier AI systems are simply not reliable enough to power fully autonomous weapons." The argument is not that autonomous weapons are inherently wrong. It is that the technology as it exists cannot be trusted to perform the way its designers intend in the messy, ambiguous, unpredictable conditions of real combat.

The concern is legitimate. Neural networks trained on millions of images perform brilliantly within the distribution of their training data. A computer vision system trained to identify T-72 tanks from overhead drone footage will identify T-72 tanks from overhead drone footage with remarkable accuracy, in clear weather, at standard altitudes, against standard backgrounds. But what happens when the weather is poor? When the tank is partially concealed by camouflage netting? When the image is degraded by rain, dust, or electronic interference? When the object in the frame is not a tank but a combine harvester that, from 500 meters up, presents a similar thermal signature? The system does not know what it does not know. It classifies every input with the same mechanical confidence, whether the classification is correct or catastrophically wrong.

This is the black box problem applied to lethal decision-making. The neural networks that power computer vision, target identification, and autonomous navigation are not transparent processes. They are statistical engines whose internal logic is not fully interpretable even by the engineers who built them. When a SLAM algorithm misidentifies a visual feature and the drone flies into a wall, the failure mode is a crashed aircraft. When a targeting algorithm misidentifies a civilian vehicle as a military transport, the failure mode is dead civilians. The technology that makes autonomous operation possible is the same technology whose failure modes are difficult to predict, difficult to detect, and, in the context of weapons, potentially irreversible.

The defense-tech industry's response is not to deny the problem but to argue that the comparison is wrong. The relevant comparison, they contend, is not between AI performance and perfect performance. It is between AI performance and human performance under the same conditions. A human FPV drone operator in Ukraine, sleep-deprived, stressed, operating in an electronic warfare environment, making split-second decisions from a grainy thermal feed, that human's error rate is not zero. The question is whether an AI system, operating at the edge, with all its limitations, makes fewer catastrophic errors than the human it supplements or replaces. The ACE program's results suggest that in the specific domain of air combat maneuvering, the answer is yes. Whether that translates to the far more ambiguous domain of target identification and engagement, the domain where "is that a tank or a tractor?" becomes "is that a combatant or a civilian?": remains an open question. It is, in many ways, the open question of this book.

From Brain to Network

Brandon Tseng's original problem (how to build a machine that operates when every link to the outside world is severed) has been solved. Not perfectly. Not without limitations. But solved well enough that Hivemind flew US Navy combat aircraft autonomously in December 2025, that Skydio's Autonomy Engine navigates GPS-denied environments at more than one million 3D points per second, and that Ukrainian FPV drones complete strike missions after losing all contact with their operators. Edge AI is not a future capability. It is a deployed reality.

But a single smart drone, no matter how capable its onboard intelligence, is still a single drone. The Nova that mapped a building in Afghanistan was alone. The FPV drone that struck its target after being jammed was alone. The X-62A VISTA that outfought a human pilot was alone. The revolution described in Part I of this book, the revolution of mass, of swarms, of hundreds of autonomous weapons overwhelming defenses designed for individual threats, requires something more than intelligence at the edge. It requires those intelligent edges to connect. A hundred drones, each running its own Hivemind or its own autonomy engine, each processing its own sensor data and making its own decisions, that is a hundred independent actors, not a coordinated force.

The transformation from a hundred individual brains into a single coordinated intelligence (a mesh of autonomous systems sharing data, distributing tasks, and executing missions as a unified formation) is the next layer of the technology stack. And it is the layer where the control problem becomes most acute. When a single drone makes an error, one target is hit or missed. When a networked swarm makes an error, when the mesh itself carries a flawed classification or a corrupted targeting decision from one node to every other node at machine speed, the error scales with the network.

One company believes it has built the operating system that makes the mesh work. Its platform is called Lattice. Its contract with the US Army is worth $20 billion. And the question it raises (who controls the network when the network operates faster than any human can review) is the subject of the next chapter.

Chapter 6: The Mesh

PART II: THE TECHNOLOGY

On a September morning in 2020, inside a command post at White Sands Missile Range in southern New Mexico, an Air Force operator strapped on a virtual reality headset built by a company called Oculus. The headset had been designed to let teenagers play video games. The operator was using it to track a simulated Russian cruise missile.

The headset displayed a three-dimensional map of the desert floor overlaid with data streaming from Air Force radar systems and a ring of autonomous sensor towers positioned across the range. The data did not come from a single source. It came from all of them simultaneously, radar returns, electro-optical feeds, radio-frequency detections, acoustic sensors, fused in real time by software that classified each object on the map, assigned it a threat level, and presented the operator with a set of response options before the missile's simulated trajectory had covered another mile of desert. The operator tagged the incoming object as hostile. The software offered intercept solutions. The operator selected one. The simulated missile died.

The software was called Lattice. The company that built it was called Anduril Industries. And the exercise (one of the Air Force's Advanced Battle Management System (ABMS) demonstrations, designed to test whether the Pentagon could compress the time between detecting a threat and neutralizing it) was a proof of concept for something far more ambitious than shooting down a single cruise missile. It was a test of the idea that software, not hardware, would be the foundation of the next military revolution. That the most important weapon on the future battlefield would not be a drone, a fighter jet, or a missile, but the network that connected all of them into a single, AI-mediated system of systems.

The operator who ran the demonstration that morning wore an Oculus headset because Palmer Luckey, the man who had invented the Oculus Rift, sold the company to Facebook in a deal initially valued at $2 billion, and then been fired, had founded Anduril three years earlier with the explicit intention of bringing Silicon Valley's software-first approach to the defense industry. The name Anduril came from *The Lord of the Rings*, the reforged sword of Aragorn, "Flame of the West." The ambition was not subtle.

The exercise itself was modest, a handful of sensor nodes, a single simulated threat track, a desert range where the biggest variable was the wind. But the concept it demonstrated was not modest at all. ABMS was the Air Force's contribution to Joint All-Domain Command and Control (JADC2), the Pentagon's grand ambition to connect every sensor, every weapon, and every decision-maker across all military services into a single networked architecture. More than 50 companies had been selected to contribute, under a contract worth up to $950 million. The idea was to compress the kill chain (the time between detecting a threat and acting on it) from hours or minutes to seconds. Lattice was one candidate among dozens. It would not remain one among dozens for long.

Six years after that White Sands demonstration, the US Army would build its entire counter-drone architecture around Lattice, and hand Anduril a $20 billion contract to make it work.

From Garage to $60 Billion

The founding story of Anduril Industries reads like a Silicon Valley origin myth engineered for a defense audience. In June 2014, Palmer Luckey attended a retreat on Sonora Island, British Columbia, hosted by Founders Fund, the venture capital firm led by Peter Thiel that had been an early Oculus investor. There, Luckey met Trae Stephens, a former Palantir employee who had recently joined the fund. The two discovered a shared frustration: aside from Palantir and SpaceX, virtually no venture-backed companies worked closely with the US government, despite the government spending hundreds of billions of dollars annually on technology. Stephens found it absurd. His goal, he later told journalists, was to fund a company to join that duo.

Three years later, after Luckey departed Facebook in March 2017 (alleging he had been fired for his political beliefs, which Facebook denied) the two incorporated Anduril Industries with three additional co-founders: Matt Grimm, Joe Chen, and Brian Schimpf. Schimpf had been a director at Palantir. When Luckey, Stephens, and Grimm pitched their defense startup concept to Palantir's leadership, Schimpf decided to leave and join them. He became CEO. Founders Fund provided the seed investment.

The plan was to apply Luckey's Oculus development philosophy, combine low-cost commercial hardware components with sophisticated proprietary software: to military systems. "The defense industry has been stagnant for decades," Luckey told reporters. The confidence was characteristic. So was the underlying bet: that the Pentagon's problem was not a shortage of hardware but a shortage of software capable of making that hardware work together.

The bet paid off faster than anyone outside the company expected. Anduril's first product line was not a weapon. It was a surveillance system, and the customer was not the Pentagon but US Customs and Border Protection. Anduril deployed autonomous Sentry Towers along the US-Mexico border: 33-foot solar-powered surveillance stations equipped with cameras, radar, thermal imaging, and communications antennas, all feeding data into Lattice. When disassembled, a Sentry Tower fit into a pickup truck. CBP agents could reassemble one in under two hours. By 2023, 290 towers were operational, forming what CBP officials described as a "virtual border wall."

The technology was not a wall at all. It was a network. Each tower operated autonomously, detecting and classifying objects in its sensor range, then passing that data to every other tower and to human operators who could see the entire border zone in a single fused picture on a laptop or phone. A person crossing the desert at night triggered a thermal alert on one tower, which was corroborated by radar on the next tower, which updated Lattice, which presented a border patrol agent with a classified track and a recommended intercept point, all before the person had walked another hundred meters. The value was not in any single tower. The value was in what the towers knew together.

From border surveillance to battlefield command. The distance was shorter than it appeared. The same software that tracked a migrant in the Sonoran Desert could track a cruise missile over New Mexico. The sensor inputs were different. The classification algorithms were different. The core architecture, ingest, fuse, classify, recommend, was identical. Anduril had built its operating system in a low-stakes environment and was now scaling it to the highest-stakes environment on earth.

The Operating System for War

What Lattice actually does is deceptively simple to describe and extraordinarily difficult to build. It is an AI-powered software platform, Anduril calls it Lattice OS, that ingests data from distributed sensors across all domains (air, ground, sea, space, cyber), fuses that data into a single real-time operational picture, and manages autonomous systems through what Anduril terms its "autonomy layer." Radar, acoustic sensors, radio-frequency detectors, electro-optical cameras, thermal imagers, and the sensor suites of autonomous platforms like Anduril's Ghost drone all feed into Lattice. The software classifies each detected object, tracks its movement, and, when a threat is identified, presents human operators with response options or coordinates autonomous effectors to neutralize the target. Military clients access the system using a laptop or a phone. The interface is designed to be usable by a soldier in the field, not just an analyst in a command post.

The Ghost drone illustrates how Lattice transforms individual platforms into network nodes. The Ghost is a small unmanned aircraft system equipped with machine learning and computer vision algorithms for target identification and tracking. It uses NVIDIA processing units, originally designed for self-driving cars, to process imagery independently, without requiring a centralized analysis system. Multiple Ghost drones can link together through Lattice to form swarms, relaying data among themselves to extend the effective communication range back to a Lattice station. US Army soldiers flew the Ghost X variant during a National Training Center rotation in November 2025. The drone had undergone substantial redesign based on feedback from Ukrainian combat operations, a feedback loop between battlefield reality and Silicon Valley engineering that traditional defense primes had never established at this speed.

The architecture matters because it solves a problem that has bedeviled the US military for decades: interoperability. The F-22 Raptor and the F-35 Lightning II, America's two premier stealth fighters, were designed with incompatible tactical data links. They cannot share targeting data directly. An Army radar and a Navy sensor speak different digital languages. A Marine Corps drone and an Air Force satellite relay information through separate networks that do not talk to each other. The Pentagon has spent billions trying to stitch these systems together under programs with ambitious names, Joint All-Domain Command and Control (JADC2), ABMS, and the progress has been incremental at best.

Lattice approaches the problem from the opposite direction. Instead of trying to make legacy systems interoperable (a task that requires rewriting decades of bespoke software across dozens of platforms) Lattice sits on top of everything, ingesting data from any sensor that can output it and translating it into a common operational picture. The platform does not care whether the data comes from an Anduril product or a third-party system. It cares about what the data means. A radar return from an Air Force installation and an acoustic detection from a Navy vessel become, inside Lattice, a single fused track with a classification, a trajectory, and a threat assessment. The operator sees one picture. The picture is updated in real time. The system recommends actions.

Consider what this means in practice. An Army counter-drone unit detects an incoming threat with ground-based radar. Simultaneously, an Anduril Ghost drone on patrol picks up the same object on its electro-optical camera. A Navy ship offshore registers an RF emission from the same bearing. In the pre-Lattice world, these three detections exist in three separate systems, managed by three separate services, interpreted by three separate analysts who may or may not be in communication. By the time the detections are correlated and a response is authorized, the threat has arrived. In the Lattice world, the three detections are fused automatically into a single track within seconds. The system classifies the threat, calculates its trajectory, and presents the nearest operator, regardless of service branch, with response options. The Anvil interceptor drone, already airborne and connected to the mesh, can be tasked to intercept before the Army analyst has finished reading the radar return.

Speed is the product. Interoperability is the method. Software is the foundation.

That is the software layer. The networking layer is what makes it a mesh.

What a Mesh Actually Is

A mesh network is an architecture in which every node communicates with every other node. There is no central hub. No single base station that all traffic must pass through. If one node is destroyed, the remaining nodes route data around the gap. The network degrades gracefully rather than failing catastrophically.

In the context of drone swarms, this means each drone acts as both a receiver and a transmitter. Data flows dynamically from drone to drone without dependence on a ground station. A reconnaissance drone detects a target and shares that data not by sending it back to base for processing and redistribution, but by passing it directly to the strike drone next to it, which passes it to the drone next to that one, which relays it to the command post three kilometers away. Every node extends the network's range. Every node that survives keeps the network alive.

Consider the difference between a traditional drone operation and a mesh-networked one. In a traditional operation, 10 drones communicate with a single ground station through 10 individual links. Destroy the ground station and all 10 drones lose their connection. Jam the frequency between the ground station and the drones and the entire fleet goes blind. The architecture has a single point of failure, and that single point is the most obvious target on the battlefield.

In a mesh architecture, each of those 10 drones communicates with every other drone. That is not 10 links: it is 45. Data can travel from any drone to any other drone through multiple paths. If Drone 3 is destroyed, Drones 2 and 4 route around the gap. If the ground station is jammed, the drones continue operating on pre-loaded mission parameters, sharing sensor data among themselves, making collective decisions about target allocation and flight paths. The network does not depend on any single node. It depends on the network itself.

Lattice Mesh is Anduril's implementation of this architecture for military operations. It distributes data securely across different military services and domains (Army sensors sharing with Air Force effectors, Navy radar feeding Marine Corps targeting) without requiring a centralized server that an adversary could locate and destroy. The mesh is the connective tissue that turns individual platforms into a system, individual sensors into awareness, individual weapons into coordinated lethality.

The concept is straightforward. The engineering is not. A mesh network that works in a controlled environment, a technology demonstration, a corporate campus, a border zone where the primary threat is a person walking through the desert, is fundamentally different from a mesh network that works on a battlefield where the electromagnetic spectrum is under active attack. Russian electronic warfare units in eastern Ukraine have demonstrated the ability to jam GPS signals across hundreds of square kilometers, disrupt radio links, and spoof drone navigation systems. Any mesh network that depends on uncontested communications is a mesh network that will fail in the one environment where it matters most.

That is why the most important advances in mesh networking are not about how the network works when everything goes right. They are about what the network does when everything goes wrong.

The Swarm That Cannot Be Decapitated

In February 2025, L3Harris Technologies unveiled a command-and-control system called AMORPHOUS. The name was not an accident. The system was designed to have no fixed shape, no single node whose destruction would collapse the swarm.

AMORPHOUS is built on a principle that inverts traditional military command architecture. In a conventional drone operation, a ground control station serves as the brain. It issues commands, receives data, and coordinates the actions of every drone in the fleet. Destroy the control station and you destroy the operation. AMORPHOUS eliminates the single point of failure by distributing command authority across every drone in the network. If the drone currently serving as the coordinator is shot down, jammed, or otherwise lost, any other drone in the swarm can autonomously assume coordination duties and continue the mission. The swarm does not pause to reorganize. It reorganizes in flight.

The technical foundation for this kind of resilience is a class of algorithms called consensus protocols. The SwarnRaft system, demonstrated in 2025, uses the Raft consensus algorithm (originally developed for distributed computer networks) to achieve state agreement across a drone swarm even when satellite navigation is entirely unavailable. Each drone maintains its own estimate of the swarm's shared state: the positions of other drones, the location of targets, the status of the mission. When drones communicate, they reconcile their estimates. When communication is intermittent or degraded, each drone falls back on its own sensors and its own pre-loaded mission intent, continuing to operate based on the last known consensus until the mesh reconnects.

The cognitive radio layer adds another dimension of resilience. Each drone in a mesh-networked swarm carries a software-defined radio that uses AI algorithms to continuously evaluate channel quality and electronic interference in real time. If the radio detects jamming on one frequency, it switches to another. If all primary frequencies are compromised, it adjusts power levels, narrows its beam, or reroutes data through alternate drone nodes that are outside the jammed zone. The radio does not wait for a human operator to diagnose the problem and select a new frequency. It adapts automatically, at machine speed, because the electromagnetic environment on a modern battlefield changes faster than any human can track.

The result is a network architecture that is designed from the ground up for degradation. Not designed to prevent degradation: that would be naive. Designed to absorb it. A mesh swarm operating in a heavily jammed environment will lose drones, lose communication links, lose GPS, lose contact with its operators. It will not, if the architecture works as designed, lose coherence. Each surviving node carries enough intelligence, enough autonomy, and enough communication flexibility to maintain the mission until the mesh can heal itself.

That "if" is the crux of the entire chapter's argument.

The $20 Billion Bet

On a Monday in March 2026, the US Army announced the largest counter-drone contract in its history. The recipient was Anduril Industries. The ceiling was $20 billion over 10 years.

The contract's structure was as significant as its dollar figure. It consolidated more than 120 separate procurement actions, individual contracts for sensors, effectors, software, integration, testing, and support that had previously been scattered across dozens of vendors and program offices, into a single enterprise agreement built around Lattice. The initial task order was approximately $87 million. But the architecture of the deal signaled something the defense establishment had been slow to acknowledge: the Pentagon was no longer buying individual counter-drone weapons. It was buying the software platform that would connect all of them.

The same week, Anduril closed a $4 billion funding round led by Andreessen Horowitz and Thrive Capital, pushing the company's valuation to $60 billion. Nine years earlier, Palmer Luckey and Trae Stephens had incorporated a defense startup seeded by a venture capital firm. Now their company was worth more than most of the legacy defense primes had been worth a decade ago, and the US Army was betting its counter-drone future on their software. The trajectory was staggering even by Silicon Valley standards: zero to $60 billion in nine years, with revenue reaching $2 billion in 2025 and a workforce of 7,000 employees. The company was simultaneously building a $1 billion hyperscale manufacturing facility in Pickaway County, Ohio, Arsenal-1, designed to produce autonomous aerial and maritime drones natively equipped with Lattice software at a pace that legacy manufacturing could not match.

The contract made concrete what the White Sands demonstration had only suggested. The most valuable technology in autonomous warfare is not the drone. It is not the sensor. It is not the missile. It is the mesh, the software layer that connects drones, sensors, missiles, and human operators into a network that sees more, reacts faster, and coordinates more precisely than any collection of individual systems could achieve on its own.

Here is the corollary that keeps defense strategists awake: whoever controls the mesh controls the battlefield. Not because the mesh fires the weapon—it does not. But because the mesh determines what the weapon sees, what it knows, how fast it can act, and whether it can coordinate with every other weapon in the network. The mesh is the nervous system. The drones and missiles and sensors are the limbs. A military with superior limbs but an inferior nervous system will lose to a military that has wired itself together more completely, more resiliently, and more intelligently. The $20 billion is not buying drones. It is buying the nervous system.

The Intellectual Architect

Christian Brose did not set out to work for a defense startup. He set out to fix the Pentagon from the inside.

As a young Senate staffer, Brose became a senior advisor to Senator John McCain, the Arizona Republican who chaired the Senate Armed Services Committee and who regarded the US military's procurement system with something between exasperation and fury. Brose rose to become the youngest-ever Staff Director of the Committee, the most powerful staff position in Congress on matters of defense policy. From that perch, he watched the US military spend trillions of dollars on exquisite platforms that took decades to develop, arrived over budget and behind schedule, and were designed for a world that no longer existed.

In 2020, Brose published *The Kill Chain: Defending America in the Future of High-Tech Warfare*. The book's thesis could be stated in a single sentence: America's military kill chain, the detect-decide-act cycle that determines whether a weapon hits its target, was broken because the Pentagon kept building expensive platforms instead of networked, AI-enabled systems, and if the country did not fix it, it would lose the next war. The argument was not polite. Brose described a military that had spent decades optimizing for wars it had already won while its adversaries designed systems to exploit precisely the weaknesses that optimization had created. The bureaucracy was the problem. The procurement system was the problem. The institutional preference for billion-dollar platforms over distributed, attritable networks was the problem. The book made the Marine Corps Commandant's reading list. It appeared on the Defense Acquisition University's professional reading list. The phrase "kill chain" entered the defense vocabulary as shorthand for the entire diagnosis. A RAND review called it "blessedly devoid of heavy theory and jargon." Military reading lists across the Navy, the Marine Corps, and the Army added it to their recommended shelves.

Then Brose did something unusual. He left government and joined Anduril Industries as Chief of Strategy. The company he joined was building exactly the kind of networked, software-defined, autonomous systems he had argued the Pentagon needed. Reviewers later noted that *The Kill Chain* had functioned as both a policy argument and, in retrospect, a business manifesto for the company Brose would join. The diagnosis and the prescription had found each other.

Brose's presence at Anduril matters for the argument of this book because it illustrates a pattern that recurs across the defense-tech ecosystem: the people who understand the problem best are leaving government to build the solutions in the private sector. This is partly a story of opportunity, defense-tech startups offer equity, speed of execution, and professional autonomy that government service cannot match. But it is also a story of frustration. The kill chain Brose diagnosed from inside the Senate remains broken. The procurement system he criticized still takes years to field systems that adversaries can build in months. The difference is that Brose is now building the fix on the outside, with venture capital backing and a $20 billion Army contract, rather than arguing for it on the inside with a white paper and a hearing.

The pattern extends beyond Brose. Anduril's co-founder Trae Stephens served on the Defense transition team and the Defense Innovation Board before leaving for the private sector. He became Anduril's chairman. Shield AI's Brandon Tseng left the Navy SEALs to build the autonomous pilot he wished he had in Afghanistan. The Defense Innovation Unit, created to bridge Silicon Valley and the Pentagon, has seen its alumni scatter across the defense-tech startup ecosystem. The talent pipeline flows in one direction: out of government, into companies that promise to move faster. Whether this represents healthy innovation or a dangerous hollowing-out of institutional knowledge is a question the defense establishment has not yet answered.

The Emerging Duopoly

The Anduril story cannot be told in isolation. The company's rise is entangled with another firm that shares its DNA: Palantir Technologies.

The entanglement is literal. Anduril's co-founder Trae Stephens was a former Palantir employee. CEO Brian Schimpf was a Palantir director. The two companies share a founding investor in Peter Thiel's Founders Fund. They share a conviction that software, not hardware, is the decisive layer in modern defense. And they increasingly share the Pentagon's AI procurement budget. In 2025, Palantir's government revenue reached approximately $2.4 billion, with defense representing the majority. Anduril's revenue reached approximately $2.15 billion. Together, two companies that did not exist 20 years ago captured a combined defense revenue rivaling legacy primes that had been building weapons since World War II.

The division of labor is becoming clear. Palantir provides the intelligence and analytics layer, its Maven Smart System, the successor to Project Maven, has been deployed in operations spanning Iran, Iraq, Syria, Ukraine, and Yemen, fusing data from satellites, military sensors, social media, and commercial data brokers to identify targets and simulate battlefield scenarios. Anduril provides the autonomous systems and command-and-control layer, Lattice, Ghost drones, Altius loitering munitions, Anvil interceptors, and the Fury CCA prototype. Palantir makes the picture. Anduril acts on it.

The convergence of these two companies into a de facto defense-tech duopoly raises questions that extend beyond market competition. In 2025, the US Army warned that a battlefield communications system co-designed by Palantir and Anduril operated as a "black box", the military could not verify whether unauthorized users could access the system's applications and data. The Army reported that it had since mitigated the issue. But the episode exposed a structural tension at the heart of the software-defined defense revolution: when the most critical layer of military infrastructure is proprietary commercial software, the institution that depends on it cannot fully inspect what is inside.

The mesh connects everything. The question is who owns the mesh, and who can see through it.

The parallel to the broader argument of this book is direct. On the Hunter-Killer spectrum, Palantir occupies the hunter end, finding, identifying, classifying. Anduril occupies the mesh that connects hunter to killer and, increasingly, the killer end itself. Together, they are building the infrastructure that compresses the Gap, the space between sensing a target and striking it where human judgment is supposed to live. When the intelligence layer and the action layer are built by allied companies running on compatible software, the compression accelerates. The kill chain that Brose diagnosed as broken is being rebuilt, not by the Pentagon, but by two venture-backed companies whose combined defense revenue now exceeds $1.8 billion per year.

The Gap in the Mesh

The honest assessment of mesh networking in 2026 is that the architecture is sound and the implementation is incomplete. Anduril's Lattice has been demonstrated at White Sands, deployed on the border, selected by the Space Force, and chosen as the backbone of the Army's counter-drone program. L3Harris AMORPHOUS has demonstrated swarm resilience in controlled conditions. Cognitive radio systems have shown the ability to adapt to jamming in laboratory and field environments. The SwarnRaft consensus protocol has achieved distributed state agreement without GPS.

None of these systems has been tested at scale in the electromagnetic hellscape of a peer-to-peer conflict.

The progress is real. In March 2025, at the Arctic Strike exercise in northern Sweden, Saab demonstrated a system that allowed a single operator to control 100 drones simultaneously, performing ISR and creating redundant communication links in a contested electronic warfare environment. The system had been developed in 12 months. Personnel could be trained to operate it in days, not months. That is the speed at which the mesh is advancing, fast enough to deliver new capabilities in the time it takes the Pentagon to draft a requirements document.

But the gap between demonstration and combat (the same gap that swallowed the Altius in eastern Ukraine) applies to mesh networking with particular force. A mesh that routes around a single lost node in a clean spectrum is not the same as a mesh that maintains coherence when dozens of nodes are destroyed simultaneously, when every frequency is jammed, when GPS is spoofed, when the adversary is deploying its own AI to attack the mesh itself. Ukraine has tested swarms of eight to 25 drones in over 100 operations. True large-scale mesh warfare (hundreds of networked autonomous systems coordinating across a contested battlespace) remains a projection, not a demonstrated capability.

The $20 billion Anduril contract is a bet that the projection will become reality. The L3Harris AMORPHOUS system is a bet that the mesh can be made resilient enough to survive what an adversary will throw at it. The cognitive radio layer is a bet that AI-driven frequency management can outrun electronic warfare. Saab's 100-drone demonstration is a bet that the ratio of operators to drones can be compressed from one-to-one to one-to-many without losing operational effectiveness. These are informed bets, backed by substantial engineering and operational evidence. They are not certainties. And the adversary gets a vote. Russia's electronic warfare capabilities in eastern Ukraine have already demonstrated the ability to degrade drone operations at a scale and sophistication that no Western test range can replicate. China's own investments in AI-enabled mesh networking, including the CETC swarm systems and the 200-drone AI swarm demonstrated in 2025, mean that the mesh will face not just jamming but competing meshes, adversarial AI attempting to penetrate, confuse, or co-opt the network itself.

The mesh is the most consequential technology in autonomous warfare because it is the layer that transforms individual systems into a collective, a swarm that is greater than the sum of its parts. It is also the layer where the control problem becomes most acute. When the mesh makes decisions faster than any human can review them, routing data, reallocating targets, coordinating strikes across dozens of platforms in milliseconds, the question is not whether a human is "in the loop." The question is whether the loop has any room left for a human to occupy.

Lattice connects the mesh. Hivemind flies the drone. But the system that makes all of it militarily useful is the one that watches, persistently, autonomously, around the clock. The next chapter examines the surveillance architecture that never blinks, and why the most important drone contract in Army history was not for a weapon but for an eye.

Chapter 7: The Eyes That Never Blink

PART II: THE TECHNOLOGY

The box sat on a concrete slab at the perimeter of a NATO forward operating base in eastern Europe, 2026. It was roughly the size of a residential air conditioning unit, gray, weatherproof, unremarkable. A soldier walking past it might have mistaken it for an equipment locker or a generator housing. It was neither. Inside the box, a Skydio X10D rested on a charging cradle, its four folded arms tucked against a body barely larger than a hardcover book. The drone had landed eleven minutes earlier, after a 35-minute autonomous patrol of the base's northern approach. Its 48-megapixel telephoto camera and Teledyne FLIR Boson+ thermal sensor (the first thermal imager of its class ever integrated into a small UAS) had surveyed 2.3 square kilometers of terrain, logging every thermal signature, every vehicle movement, every anomaly that deviated from the baseline pattern established over the previous 72 hours of continuous coverage.

At 0247 local time, a sensor alert triggered. The Dock's automated launch sequence activated. The lid opened. The X10D unfolded its arms, spun its rotors, and lifted into the night air. Twenty seconds from alert to airborne. No human touched a controller. No pilot rubbed sleep from their eyes or fumbled for goggles. The drone climbed to 80 meters and navigated autonomously toward the alert coordinates using its six navigation cameras, three above, three below, 360-degree awareness with no blind spots, building a real-time 3D map of the terrain through Visual-Inertial Odometry, the same GPS-denied navigation technology that had been validated in the electronic warfare hellscape of eastern Ukraine. It did not need satellites. It did not need a radio link to a ground station. It saw the world and knew where it was.

Fourteen hundred meters south, in the base's tactical operations center, a sergeant monitoring a browser-based interface saw the drone's feed appear on her screen. The thermal camera had acquired two heat signatures moving along a drainage channel that ran parallel to the perimeter fence. The AI flagged the signatures as anomalous, the pattern did not match any scheduled patrol or known wildlife track. The sergeant alerted the quick reaction force. By the time the QRF reached the channel, the X10D had been tracking the signatures for four minutes, relaying continuous video through Skydio's multiband radio, its dynamic frequency switching keeping the data link alive in a spectrum environment where Russian-supplied jammers routinely degraded commercial frequencies. When the X10D's battery dropped below 15%, it broke contact, returned to the Dock, and landed. A second drone from an adjacent Dock, 200 meters east, was already airborne, picking up the surveillance pattern within seconds of the first drone's departure.

This is what Skydio calls a Hive. Two or more automated Docks, networked together, executing seamless handoffs between drones so that coverage never lapses. One drone charges while another flies. The cycle repeats, hour after hour, day after day, without human fatigue, without attention gaps, without the 3 a.m. lapse in concentration that has been the bane of perimeter security since the invention of the guard tower. The military term for what the Hive provides is persistent ISR, Intelligence, Surveillance, and Reconnaissance that does not stop. The strategic implications of that persistence are what make this chapter necessary, even though Skydio's founding, technology, and defense pivot were covered in Chapter 2. The earlier chapter explained the hunter. This one explains the architecture that lets the hunter watch forever.

The Dock That Never Sleeps

The Skydio Dock is the hardware that makes persistence possible. Stripped to its essentials, it is an automated base station, a weatherproof housing with a charging cradle, a launch mechanism, and the connectivity to receive mission commands from anywhere on the network. The Dock manages every phase of the drone's operational cycle: automated launch, automated landing, automated recharging. A single off-site operator can oversee the entire system from a browser. Or the system can run fully autonomously, responding to sensor alerts, executing pre-programmed patrol routes, and cycling drones through their missions without any human input at all.

The numbers that matter are small but consequential. Under 20 seconds from rest to airborne. That response time transforms the Dock from a convenience into a tactical asset. A traditional drone operation requires a pilot to be awake, alert, and physically present with the aircraft. The pilot must unpack the drone, power it up, establish a control link, and fly it to the area of interest, a process that takes minutes in ideal conditions and longer in the dark, the cold, or under fire. The Dock compresses that sequence to a single automated action. An alert comes in. The drone launches. Twenty seconds later, it is over the target.

The Hive architecture multiplies this capability. When two or more Docks are networked (placed at strategic intervals around a base perimeter, a supply route, a border crossing, a critical infrastructure site) they form a system that provides continuous coverage without requiring continuous human attention. The Docks communicate with each other and with Skydio's Remote Operations platform, which handles fleet management, mission planning, and the automatic deconfliction of airspace when multiple drones are operating simultaneously. The handoff between drones is seamless: as one drone returns to its Dock to recharge, the next is already airborne, maintaining the surveillance pattern without interruption.

This architecture resolves a constraint that has limited military drone operations since their inception: the one-to-one ratio between operator and aircraft. Traditional drone operations require one trained human per drone. A Reaper requires a crew of at least two. Even a small tactical quadcopter demands a pilot's full attention for the duration of the flight. The Hive inverts that ratio. Skydio CEO Adam Bry has described the vision as the "1:many" model, one operator overseeing five, ten, or fifty drones, each flying autonomously while the human provides intent and oversight, not stick-and-rudder control. "The drone should work for the person, not the other way around," Bry has said. "And the only way you are going to get there is with AI and autonomy, making the thing smart enough to fly itself."

The 1:many model is not a projection. It is operational. And its most advanced deployment is not on a battlefield. It is in American cities.

Ten Million Calls

In February 2026, Skydio announced that its Drone as First Responder (DFR) Command platform had surpassed 10 million calls for service.

That number deserves a pause. Ten million is not a technology demonstration. It is not a pilot program. It is operational scale, the kind of number that proves a system works not in a controlled environment but in the unpredictable, high-stakes, legally scrutinized world of American law enforcement. DFR Command is the software layer that combines Skydio's Dock hardware, Remote Operations platform, and Autonomy Engine into a complete system for public safety agencies. A 911 call comes in. A gunshot detection system flags an alert. The nearest Dock launches a drone. The drone arrives on scene in under 60 seconds (some agencies report response times significantly faster than that) and provides live aerial video to dispatchers and responding officers before any patrol car arrives.

The integrations tell the story of how deeply the system has penetrated the public safety ecosystem. DFR Command connects with more than 25 different public safety platforms. Its integration with SoundThinking (the acoustic gunshot detection network formerly known as ShotSpotter) allows drones to auto-launch the moment a gunshot is detected, without any human initiating the flight. The system hears the shot, calculates the location, and dispatches the nearest available drone. By the time officers arrive, they have aerial footage of the scene, the surrounding streets, and any fleeing suspects.

For defense planners, the DFR numbers are more than a commercial success story. They are a proof of concept for the persistent autonomous surveillance architecture that militaries are now deploying. The same Dock that launches a drone to a gunshot alert in a midwestern city is the Dock that launches a drone to a perimeter breach on a NATO base. The same AI that navigates between buildings at night without GPS in an urban environment is the AI that navigates between tree lines in a contested border zone. The same seamless handoff that keeps a neighborhood under continuous aerial observation keeps a supply route under continuous aerial observation. Skydio built the system for police departments. The military implications came free.

The convergence between domestic law enforcement and military surveillance technology is not coincidental, and it is not without critics. A system that watches a neighborhood 24 hours a day raises civil liberties questions that a system watching a battlefield perimeter does not. But the technical architecture is identical, and the 10 million calls for service provide something that military testing ranges cannot: statistical proof that the system works at scale, in real-world conditions, over extended periods of time.

The X10D at War

While DFR Command proved the architecture in American cities, Ukraine proved it in combat.

Since 2022, Skydio has delivered approximately 1,000 drones to the Ukrainian Defense Forces. Skydio personnel made more than 30 visits to the combat zone over a two-year period, an extraordinary level of engagement for an American technology company in an active war. The visits were not goodwill tours. They were engineering sprints. Skydio teams worked directly with Ukrainian operators, observed how the drones performed under Russian electronic warfare, identified failure modes that no stateside testing range could replicate, and fed that data directly to the engineering team in California. The X10D was, in Skydio's own phrasing, "sharpened by hard lessons in Ukraine and close iterative feedback with soldiers."

The operational pattern that emerged in Ukraine was the hunter-killer pairing, the tactical architecture that now defines small-drone warfare on the eastern front. A Skydio X10D, or a similar ISR platform, identifies and locates a target. It feeds the target's coordinates into Ukraine's Delta battlefield management system, a digital command layer that connects sensors to shooters across the front. Within minutes, sometimes within seconds, a strike drone follows: an FPV drone carrying a shaped charge, or a loitering munition waiting in a holding pattern for a target assignment. The hunter finds. The killer destroys. Two drones, two functions, one kill chain.

The X10D's role in this pairing is pure sensing. It does not carry weapons. It does not strike targets. But without the hunter, the killer is blind, a fact that Ukrainian operators understood with a clarity born from thousands of engagements. The hunter is not the dramatic half of the pairing. There are no viral videos of an X10D streaking toward a target and detonating. Its contribution is invisible to the audience that consumes combat footage on social media. But ask any Ukrainian drone unit commander which half of the hunter-killer pair is more valuable, and the answer is consistent: the one that finds the target. You can always find another $200 FPV drone. Finding the target in a contested, GPS-denied, EW-saturated environment: that is the hard part.

Bry recognized this during his visits. "I have never met drone users as sophisticated as the folks in Ukraine," he said in December 2025. "We believe it is where the future of defense is being shaped." The statement was not false modesty. Ukrainian operators had been flying drones in combat for nearly three years by then, longer, with more intensity, and against a more capable adversary than any other military in the world. They knew what worked and what did not. Their feedback carried the authority of experience that no amount of simulated testing could substitute.

The lessons that flowed from Ukraine to California were specific and consequential. GPS-denied navigation, the capability that Bry and Bachrach had developed in an MIT parking garage and refined over a decade of engineering, was not a theoretical advantage. It was a survival requirement. Russian electronic warfare units had demonstrated they could degrade satellite navigation signals across hundreds of square kilometers. A drone that depended on GPS was a drone that would lose its position, drift, and crash, or, worse, be spoofed into flying to a location chosen by the enemy. The X10D's Visual-Inertial Odometry, which tracked position using nothing but its cameras and an onboard inertial sensor, was the capability that kept the drone operational when everything else failed. Ukraine did not validate Skydio's technology in the abstract. It validated it under conditions specifically designed to destroy it.

The American Drone Market After DJI

On December 22, 2025, the Federal Communications Commission added all new foreign-made unmanned aircraft systems and UAS critical components to its Covered List, based on a national security determination that these products "pose unacceptable risks to the national security of the United States."

The action did not arrive without warning. The Countering CCP Drones Act (legislation that would have banned DJI drones outright) had been proposed and debated for years but never passed as standalone legislation. The 2025 National Defense

Authorization Act took a different approach: it inserted a review trigger. If no federal agency completed a formal security audit of DJI by December 23, 2025, DJI's products would be automatically added to the FCC Covered List. No audit was completed. The trigger tripped. But instead of targeting DJI alone, the FCC banned all new foreign-made drones from receiving equipment authorization, a scope broader than most industry observers had expected.

The practical effect was precise. Without FCC authorization, no new foreign drone model could be imported, marketed, or sold in the United States. The word "new" mattered. Existing DJI drones that had already received FCC approval continued to operate legally. The ban did not ground a single aircraft. What it did was close the pipeline for future DJI products, no new Mavic models, no new Matrice platforms, no next-generation systems. For a company that controls more than 70% of the global consumer drone market, the closure of the American market for new products was a structural blow.

For the American drone industry, the effects were immediate and polarized. On one side, Skydio, Brinc Drones, and Unusual Machines stood to inherit the domestic market that DJI could no longer supply. On the other, the commercial drone operators who had built their businesses on DJI hardware faced an existential problem. Surveys conducted after the ban found that 43% of commercial drone pilots anticipated "extremely negative" or "business-ending" impacts. More than 80% predicted they would be out of business within two years without access to new DJI products. The scale of DJI's market dominance meant that there was no painless transition.

Skydio positioned itself as the answer. "The first and only company manufacturing small autonomous drones at scale in the U.S.", the tagline appeared in press releases, Congressional testimony, and trade show presentations. Bry had testified before the House Select Committee on the Chinese Communist Party in June 2024, making the national security case against Chinese drone dependence. He argued that DJI drones transmitted telemetry data through servers accessible to the Chinese government, that Beijing could use supply chains as a strategic weapon, and that the United States needed domestic manufacturing capability for a technology that had become central to both military and civilian operations. When the ban came six months after his testimony, the positioning was seamless.

The criticism was equally pointed. "Not only is your lobbying crippling existing life-saving technology," one drone industry expert said, "but also it is hindering innovation at large." Another was blunter: "Skydio is actively destroying the U.S. drone industry." The accusation was specific: that Skydio, unable to compete with DJI on price or product quality in the consumer market, had pivoted to competing through regulation instead, lobbying to legislate its primary competitor out of the American market rather than building a product that could win on its own merits. Bry's response was a flat denial. "We had nothing to do with it," he said of the ban legislation. The claim was difficult to reconcile with his Congressional testimony, Skydio's well-resourced policy operation led by Brendan Groves (a former Associate Deputy Attorney General and Air Force officer with deep national security credentials) and the company's relentless framing of Chinese drones as a security threat.

The honest assessment is that both sides hold partial truths. The national security concerns about Chinese drone data are not fabricated. DJI's data practices have been flagged by the Department of Homeland Security, the Department of Defense, and multiple independent cybersecurity audits. But the 3-to-4-times price gap between the Skydio X10 and a comparable DJI platform, $15,000-$20,000 versus roughly $5,000, is real, and it means that the American alternative is genuinely unaffordable for many of the government agencies and commercial operators that depend on drone technology. The ban may have solved a national security problem. It simultaneously created an affordability problem that the American drone industry has not yet solved.

The Cracks That Remain

Skydio's trajectory from MIT lab to NATO standard is a story that defense-tech boosters tell with relish, and the core narrative is genuine: the Autonomy Engine is a technical achievement of the first order, the Ukraine deployment proved the system in conditions no other Western small drone has faced, and the $52 million Army order in March 2026, 2,500 X10Ds, largest single-vendor tactical sUAS order in Army history, bid to award in under 72 hours, demonstrated that the Pentagon is willing to move fast when it finds a system that works.

But the full picture includes vulnerabilities that the success narrative tends to obscure.

The SRR Tranche 2 loss was the most visible setback. In November 2024, the US Army awarded the Short-Range Reconnaissance Tranche 2 Program of Record contract not to Skydio but to Red Cat Holdings, the parent company of Teal

Drones. The Black Widow drone that Red Cat offered was smaller and cheaper. The Army's rationale was explicit: competition. Rather than locking into a single vendor, the Army wanted multiple suppliers, a hedge against the kind of supply chain crisis that had already bitten Skydio once. Skydio continued to receive SRR orders alongside Red Cat, winning a $7.9 million low-rate initial production contract in October 2025. But losing the primary program of record to a smaller, less proven competitor was a wound that signaled the limits of technical superiority when price and supply chain resilience are factored in.

The supply chain crisis itself had been painful and illustrative. Despite marketing itself as an American manufacturer, Skydio had historically relied on Chinese components for critical subsystems, including, fatally, lithium-ion batteries from Dongguan Poweramp, a TDK subsidiary. In October 2024, China imposed sanctions that blocked Skydio's access to its sole battery supplier. The company was forced to ration batteries to one per drone, a limitation that degraded the very persistence that was Skydio's operational selling point. A drone that watches forever is less impressive when it ships with one battery. In March 2025, Beijing placed Skydio on its export control list alongside 14 other US entities. By early 2026, Skydio was touting a "fully domestic" supply chain, but the transition had been reactive, not proactive, and the vulnerability it exposed was not unique to Skydio. China controls more than 70% of the global lithium-ion battery market. Any American drone manufacturer claiming supply chain independence must answer the question of where its batteries come from, and whether that source can survive a geopolitical disruption.

The production question loomed larger than Skydio's monthly output suggested. The Hayward facility produced more than a thousand drones per month in early 2026, an impressive rate for a company that did not exist 12 years earlier. But the Pentagon's Drone Dominance campaign, led by Defense Secretary Pete Hegseth, envisions procuring at least 30,000 small drones in the near term. At a thousand per month, filling that kind of demand would take Skydio more than two years of exclusive production, and Skydio would not be the exclusive supplier. The same scaling challenge that limits every defense-tech startup applies here: building a production line that can surge from demonstration quantities to wartime quantities requires capital, workforce expansion, and supply chain depth that venture-funded companies do not automatically possess.

The Architecture of Seeing

The criticisms are real but they do not change the strategic significance of what Skydio has built. Set aside the price gap and the supply chain questions for a moment. Look instead at the architecture.

What Skydio has assembled is not a drone. It is a persistent surveillance system whose individual components, the X10D aircraft, the Dock, the Hive network, the Remote Operations platform, the DFR Command software, the Autonomy Engine, combine into something greater than their sum. The system watches a defined area continuously, autonomously, and indefinitely. It responds to alerts faster than any human operator could be roused and briefed. It navigates without GPS in electronically contested environments. It hands off between aircraft seamlessly, so that coverage never lapses. It integrates with dispatch and command systems, 25 of them in the law enforcement context, the Delta battlefield management system in Ukraine, and NATO C2 architectures across 29 allied nations.

This architecture represents the hunter end of the Hunter-Killer spectrum scaled to persistence. A single ISR flight is a snapshot, valuable, time-limited, dependent on having the drone overhead at the right moment. Persistent autonomous surveillance is a different category of capability. It means that the battlefield is always visible, always monitored, always analyzed. It means that any change, a vehicle that was not there an hour ago, a heat signature in a building that was empty at dusk, a pattern of movement that deviates from the established baseline, is detected and flagged automatically, around the clock, without the attention lapses and fatigue that degrade human watchstanders after the fourth or fifth hour of a 12-hour shift.

The military that controls this architecture sees more, knows more, and reacts faster than any adversary relying on intermittent surveillance. Twenty-nine nations have reached the same conclusion and bought into the same system. The $52 million Army order was not a drone purchase. It was an investment in an architecture of seeing.

But that architecture raises a question that its designers have not fully answered. Persistent autonomous surveillance—the ability to watch everything, everywhere, all the time—is a capability that lives on the "hunter" end of the spectrum. It does not kill. But the distance between watching and killing is measured in software, not in miles. When a Hive network detects an anomaly, flags it as a threat, and tracks it continuously for a human analyst to evaluate, the human makes the decision. When that same system feeds its data directly into a kill chain that launches a strike drone within seconds of target identification, the human decision space compresses to almost nothing. The technology that watches forever is one software layer away from the technology that kills autonomously. The distance between the Dock that launches an ISR drone to a sensor alert and the Dock that launches a strike drone to a targeting solution is not a technical gap. It is a policy choice.

Bry is aware of this. "We are on a faster track to artificial general intelligence than much of our civilization realizes," he has said. "But even in a world with AGI, human intent, initiative, empathy, and judgment will remain essential ingredients for ethical conflict." The X10D is an ISR platform. It carries cameras, not weapons. Skydio's official position (and Bry's personal conviction, as far as his public statements reveal) is that the human remains in the decision loop. The drone finds. The human decides. The system that kills is a separate system, operated by a separate actor, authorized by a separate chain of command.

That distinction is clean in theory. In the hunter-killer pairing that Ukraine has operationalized, the gap between finding and killing is already measured in seconds. As AI takes over more of the sensing, tracking, and target-identification functions (as the hunter becomes more autonomous, more persistent, more capable) the pressure to automate the killing side grows in proportion. Not because anyone wants autonomous killing, but because the kill chain operates at a speed that human decision-making struggles to match. When the hunter never blinks, the question becomes whether the killer has to wait for a human to blink first.

The drones can see in the dark. They can navigate without satellites. They can watch without sleeping. But every one of these capabilities depends on electromagnetic signals, radio links carrying data from drone to operator, sensor feeds flowing through the spectrum, navigation cues extracted from the visual environment when GPS is denied. And on the modern battlefield, the electromagnetic spectrum is the most contested domain of all. The invisible war between drones and the systems designed to blind them is not a sideshow. It is the contest that determines whether the eyes that never blink can keep seeing, or whether they go dark.

PART II: THE TECHNOLOGY

Chapter 8: The Invisible War

On a stretch of coastal water off the southeastern United States in the fall of 2025, a Northrop Grumman engineer named nothing that would identify him in an operational report watched a screen that looked unremarkable, a laptop running Tactical Assault Kit software, the kind of ruggedized mapping interface that soldiers and special operators have used for a decade. On the screen, icons representing unmanned platforms were scattered across a 50-square-mile operational area. Some icons were drones. Some were unmanned surface vessels skimming the water at low speed. Some were ground-based robots, squat and slow, positioned on a simulated shoreline. Each platform carried the same payload: a device called Rock Ridge.

Rock Ridge was an electronic warfare transceiver built by Northrop Grumman under a program called Tactical Edge Electromagnetic Solutions, TEEMS. It was packaged into a 1U modular payload smaller than a business card. Not smaller than a briefcase. Not smaller than a laptop. Smaller than a *business card*, the smallest and most capable electronic warfare system anyone at the demonstration had ever seen in an operational configuration. The engineer tapped a command. Across the operational area, the distributed network of Rock Ridge devices synchronized, geolocated a frequency-agile target emitter, the kind of signal that traditional EW platforms struggle to track because it hops across the spectrum faster than most systems can follow—and performed stand-in jamming. In a single coordinated action, the network simultaneously knocked out three different radios operating across a wide frequency range. Three radios. One command. Dozens of expendable platforms spread across fifty square miles.

The demonstration was part of Silent Swarm 2025, an annual US Navy exercise designed to test emerging concepts for unmanned warfare in contested electromagnetic environments. The exercise itself was not classified. The implications were.

Here is what had just happened, translated from the jargon: electronic warfare (the ability to detect, locate, jam, and destroy enemy communications and radar systems) had been liberated from the monopoly of the most expensive aircraft in the US military's inventory. For decades, the electromagnetic spectrum had been the exclusive hunting ground of platforms like the EA-18G Growler, a $60 million to $100 million electronic attack aircraft crewed by a pilot and an electronic warfare officer, each one a scarce and irreplaceable asset. The Growler jams enemy radars so that strike aircraft can penetrate defended airspace. It is indispensable, and the US Navy and Marine Corps operate only about 160 of them. Lose one, and you lose a crew, an aircraft, and a capability that takes years to replace.

Rock Ridge changed that arithmetic. An EW transceiver smaller than a business card, mounted on a drone that costs a fraction of what a Growler costs to fuel for a single sortie, had just done what once required a dedicated electronic attack squadron. Not as well. Not with the same power or the same range. But across a distributed network, with redundancy built in, with no crew to lose, and at a price point that made expendability not just acceptable but the entire point.

The Invisible War had entered a new phase.

The Spectrum Nobody Sees

Electronic warfare is the dimension of modern combat that the public almost never thinks about, and that military professionals think about constantly.

Every drone, every guided missile, every satellite navigation fix, every radio call between a platoon leader and a fire support officer, every radar pulse that tracks an incoming aircraft depends on the electromagnetic spectrum. The spectrum is not a metaphor. It is the physical medium through which modern warfare operates. Without access to specific frequencies, GPS stops working, communications collapse, targeting data cannot flow from sensor to shooter, and the entire kill chain, the detect-decide-act cycle that defines how militaries fight: goes dark.

Electronic warfare is the discipline of controlling that medium. It has three branches. Electronic attack uses energy (jamming signals, spoofing GPS coordinates, frying circuitry with directed microwave pulses) to deny the enemy access to the spectrum. Electronic protection hardens friendly systems against those same attacks. Electronic warfare support passively monitors the spectrum, mapping enemy emitters and building an electronic order of battle. Every major military on Earth maintains EW capabilities. Few discuss them publicly, because the details of what frequencies a country can jam, spoof, or intercept are among the most closely guarded secrets in defense.

For the general reader, the critical thing to understand is this: electronic warfare determines whether drones work. A drone that cannot receive GPS signals cannot navigate. A drone that cannot maintain a radio link to its operator cannot be controlled. A drone whose sensor data is jammed at the transmission point is flying blind. The invisible contest between drone operators and EW systems (the war for control of the electromagnetic spectrum) is not a sideshow to the drone revolution. It is the battlefield condition that shapes everything else. It is the environment in which the hunter must see and the killer must strike.

And in Ukraine, that environment has produced an arms race that is driving the most consequential development in autonomous weapons: the push toward drones that do not need a human at all.

The Gauntlet

The scale of Russian electronic warfare in Ukraine is difficult to overstate.

By late 2022, Russia had established EW complexes across the Donbas at concentrations of up to 10 systems per 21 kilometers of frontage, roughly one major jamming platform every two kilometers. EW posts were positioned approximately every 10 kilometers along the front line, set back about 6 kilometers from the forward edge of the battle area. The Palantin system, one of Russia's most capable platforms, could suppress satellite radio navigation signals along the entire line of contact and replace them with spoofed coordinates, GPS data that told a drone it was somewhere it was not, sending it spiraling into the ground or drifting harmlessly away from its target. The Borisoglebsk-2 added wider frequency bandwidth and faster spectrum scanning. The Layer system provided broader area suppression. By October 2023, *The Economist* reported that

Russia was installing video feedback and control jammers directly onto tanks, artillery pieces, and command vehicles, turning individual high-value platforms into their own electronic shields.

The results were devastating. By summer 2023, Ukraine was losing between 5,000 and 10,000 drones per month, roughly 160 to 300 per day, according to Ukrainian soldiers and a May 2023 report by the Royal United Services Institute. Small quadcopter drones had an average life expectancy of three flights. The TB2, the folk hero of the war's opening months, averaged six flights before being jammed, spoofed, or shot down. Electronic warfare contributed to the destruction of 90% of the thousands of drones Ukraine possessed at the start of the invasion. Only one-third of Ukrainian drone missions could be considered successful.

The airspace above the front lines was not empty. It was saturated with invisible energy, a wall of electromagnetic interference stretching across hundreds of kilometers, designed to make every frequency unusable, every GPS coordinate a lie. Russian forces built jammers small enough to fit in a backpack. Pocket-sized jammers were issued to individual soldiers. Ukrainian forces responded with their own jammers, mounted on trenches and vehicles. The spectrum had become the most contested territory in the war, more fought over, meter by meter, than any physical trench line.

Every Ukrainian drone that flew toward a Russian position entered this gauntlet. The jamming began kilometers before the target. First, the GPS signal degraded or disappeared. Then the video feed began to break apart, pixelation, freezing, the telltale signs of a control link under attack. Then the radio connection severed entirely, and the operator lost the drone. Another $200 weapon, dead before it reached its target. Another data point in a war of attrition measured not in territory but in electromagnetic dominance.

The jammers were winning. The drones kept dying.

And then the adaptations began.

The Wire and the Algorithm

The Ukrainian response came in two forms, and both pointed toward the same destination: a battlefield where human control of individual drones becomes optional.

The first adaptation was physical. Russia pioneered the deployment of FPV drones trailing fiber-optic cables, thin glass threads, 5 to 20 kilometers in length, spooled out behind the drone as it flew. The concept was borrowed from a weapon system that most military planners had assumed was a relic: the wire-guided missile, the kind of technology Cold War tank crews trained to use before laser and GPS guidance made it obsolete. A fiber-optic cable carried the control signals and video feed through a physical wire rather than a radio frequency. No radio signal meant no signal to jam. No GPS link meant no link to spoof. No electromagnetic emission meant the operator's position stayed hidden. The drone was, in the precise technical sense, unjammable.

The advantages were immediate and significant. Fiber-optic drones provided operators with high-quality, zero-latency video feeds. They could operate in locations where radio contact was poor or impossible, urban canyons, deep valleys, areas saturated with jamming. They required less power for communication, which meant they could idle on the ground for hours in ambush positions, consuming almost no energy while waiting for a target. Their control latency was lower than radio-linked drones, and in practice, their range and agility often matched or exceeded their wireless counterparts.

Ukrainian forces confronting Russian fiber-optic drones for the first time had no electronic countermeasure that worked against them. The jammers that had been destroying thousands of drones per month were suddenly useless. Desperate soldiers resorted to shooting the drones with rifles and machine guns, small arms against a target the size of a dinner plate, moving at 80 kilometers per hour, trailing a nearly invisible thread of glass. Both sides adopted the technology with startling speed. By May 2025, *The Washington Post* reported that Ukraine was scrambling to overcome Russia's growing edge in fiber-optic drone technology, a remarkable reversal for the country that had been the clear innovator in drone warfare only months earlier.

The fiber-optic drone solved the jamming problem by bypassing the spectrum entirely. But it introduced its own constraints: the physical tether limited maneuverability, the cable could snag on terrain, and the maximum range was bounded by the spool. The fiber-optic drone was a clever tactical adaptation, not a permanent solution. The permanent solution required something else.

That something else was software.

In March 2024, footage surfaced on social media showing a Ukrainian FPV drone approaching a Russian position when its video feed disintegrated, the visual signature of an electronic warfare system severing the radio link. The operator lost control. The drone should have tumbled from the sky. Instead, it completed the strike. An onboard algorithm, trained to recognize the target's visual signature, guided the drone through the final seconds of flight and into the target without any human input. The operator's screen was static. The algorithm finished the job.

This was automated terminal attack: the moment when a drone no longer needs its human operator to complete the kill. The human selected the target. The human guided the approach. But in the critical final seconds (the seconds where jamming is most intense, where the enemy's EW has the best chance of severing the link) the machine took over. It locked onto the visual profile, adjusted its flight path, and drove itself to impact.

The logic of this adaptation may be the most consequential in the entire autonomous weapons revolution: every time a jammer cuts the human operator's link, the drone must become more autonomous to survive. Electronic warfare does not just degrade drone performance. It *selects for autonomy.* The drones that survive the gauntlet are the drones that can operate without a human. The drones that die are the ones that cannot. The electromagnetic spectrum, invisible and silent, is the evolutionary pressure pushing unmanned weapons toward independence from their creators.

Each step in the escalation has followed this pattern with the regularity of a physical law. Russia deployed jammers. Ukraine built drones that could complete strikes after being jammed. Russia deployed more jammers, at closer range, mounted on individual vehicles and carried by individual soldiers. Ukraine developed cognitive radio, AI-driven software-defined radios that evaluate channel quality and interference in real time, automatically switching frequencies, adjusting power, or rerouting data through alternate paths to keep the control link alive under jamming. The cognitive radio does not wait for a human operator to diagnose the interference and select a new frequency. It detects the jamming, classifies the signal, and adapts, all within milliseconds, all without human input. Russia adapted its jamming to counter the frequency-hopping. Ukraine pushed more autonomy onto the drone itself, pre-loaded mission objectives that the drone receives before launch and executes independently if communications are lost, onboard target recognition using computer vision trained on thousands of combat engagements, and the ability to continue executing the mission after all communications are severed.

The pattern is an arms race, but it is an arms race with a direction. And the direction is always the same: away from the human, toward the machine.

Swarm EW: The Growler's Successor

The implications of this arms race extend far beyond Ukraine.

For decades, the United States and its allies have conducted electronic warfare from a small number of exquisite manned platforms. The EA-18G Growler is the standard: a two-seat strike fighter modified to carry jamming pods and electronic attack equipment, flown by a pilot and an electronic warfare officer, operating from aircraft carriers or shore-based airfields. The Growler is extraordinarily capable. It can suppress enemy air defenses across a wide frequency band, escort strike packages into defended airspace, and collect signals intelligence. The US Navy and Marine Corps depend on it for every major combat operation.

The Growler is also extraordinarily expensive, and there are not enough of them. At $60 million to over $100 million per aircraft, the fleet is small, approximately 160 aircraft total. Each one is a high-value target. Each one carries a crew that represents years of specialized training. Lose a Growler and its crew in the opening hours of a conflict, and the capability gap is not something that can be filled by the next production run. The production line is measured in years, not weeks. The training pipeline is even slower.

The comparison between traditional EW and what Northrop Grumman demonstrated at Silent Swarm 2025 is the same economic inversion that defines the broader drone revolution, applied to the electromagnetic spectrum.

A Growler costs $60 million to $100 million. A TEEMS-equipped drone costs $10,000 to $500,000, depending on the platform. A Growler operates as a single-point jammer, one aircraft, one crew, one location in the sky, covering a broad frequency range from a centralized platform. A TEEMS swarm distributes that coverage across dozens or hundreds of networked nodes, creating a multi-angle jamming geometry that is far harder for an adversary to counter. A Growler relies on crew-operated, pre-planned electronic attack missions. A TEEMS network uses AI-driven cognitive radio that adapts in real time, classifying signals, selecting jamming techniques, adjusting power and frequency without waiting for a human operator to make the call. A Growler survives through standoff range and speed, staying far enough from enemy defenses that it cannot be shot down. A TEEMS drone survives through expendability, it is designed to be lost, and the swarm's capability degrades gracefully as individual nodes are destroyed because any surviving drone can autonomously assume coordination duties.

The most significant difference is the one that connects to the chapter's central argument. The Growler keeps a human in the loop at every stage of its electronic warfare mission. The EW officer selects targets, manages jamming, and makes decisions in real time. A distributed TEEMS swarm, operating across fifty square miles with dozens of autonomous nodes adapting to a contested spectrum in milliseconds, cannot function that way. The speed of electronic warfare (signals moving at the speed of light, frequency changes measured in microseconds) exceeds the speed of human cognition. An AI-driven cognitive radio that detects an enemy emitter, classifies it, selects a jamming technique, and executes the attack in the time it takes a human operator to blink is not an enhancement of human capability. It is a replacement of human decision-making in the electromagnetic domain.

The same logic that drives Ukrainian FPV drones toward autonomy when their operators are jammed drives electronic warfare swarms toward autonomy when the spectrum moves too fast for human control. The pressure comes from different directions (in Ukraine, it is the enemy's jammer; in a TEEMS swarm, it is the speed of physics itself) but the destination is the same.

The $7.5 Billion Problem

The other side of the EW revolution is defense: how do you stop the drones?

The counter-drone market was already valued at over $4 billion in 2025, and it was growing fast. The US national defense budget for fiscal year 2026 dedicates approximately $7.5 billion toward counter-unmanned aircraft systems: C-UAS in Pentagon shorthand. Within that figure, the National Defense Authorization Act specifically allocates $3.187 billion for C-UAS investments. Those numbers represent one of the fastest-growing line items in the entire defense budget, and they reflect a recognition that has been slow to arrive but is now urgent: the existing air defense architecture was not designed for the threat it faces.

The problem is architectural. Traditional air defense systems, Patriot batteries, NASAMS, even short-range systems like the Avenger, were designed to intercept a relatively small number of relatively large, relatively expensive targets: aircraft, cruise missiles, ballistic missiles. A Patriot missile costs roughly $4 million per interceptor. Using it to shoot down a $200 FPV drone is not just economically absurd. It is strategically unsustainable. The magazine depth of any missile-based system is finite. A swarm that sends 50 drones at a defended position can exhaust the interceptor supply faster than it can be replenished. The math is the same math that drives the drone revolution itself, cost asymmetry, attrition curves that favor the cheaper system, the strategic advantage of expendability.

Effective counter-swarm defense requires a layered architecture: diverse, integrated sensors to track small, low-flying drones that conventional radars often miss; multiple types of effectors, electronic jammers, directed-energy weapons, kinetic interceptors, drone-on-drone systems, arranged in concentric defensive rings that progressively thin the incoming swarm; and AI-driven battle management systems that can match specific threats to the appropriate countermeasure in a matter of seconds. No single system can defeat a coordinated drone swarm. Only a system-of-systems approach (sensors, effectors, and AI tied together through software) stands a chance.

This is why Anduril's Lattice platform and its $20 billion Army contract sit at the center of the counter-drone story. Lattice is not a weapon. It is the software that connects weapons, sensors, and decision-makers into a network that can respond at machine speed. The counter-drone problem is, at its core, a software problem: the hardware to detect and destroy small drones exists in various forms, radar, electro-optical trackers, acoustic sensors, RF detectors, but each sensor type has blind spots. A radar that tracks a fixed-wing drone at 5 kilometers may miss a hovering quadcopter at 500 meters. An acoustic sensor that hears a propeller-driven drone cannot hear a gliding one. The software that fuses these sensor feeds, classifies threats, assigns the right effector to the right target, and manages engagement across a defended area in real time is the capability that determines whether the defense holds or collapses under the weight of the swarm.

The $7.5 billion figure is the price of admission to a contest that has no foreseeable end. Every improvement in counter-drone technology provokes an adaptation in drone design. Every new jammer drives the development of more autonomous drones. Every directed-energy weapon provokes swarm tactics designed to saturate it. The offense-defense cycle in the electromagnetic spectrum is the fastest-moving arms race in military technology, and it is accelerating.

Silent Swarm 2026: The Dress Rehearsal

NATO is preparing for the next phase.

Silent Swarm 2026, planned as a multinational live-fire exercise in Estonia, is designed to showcase and test NATO doctrine for countering autonomous drone swarms in a contested electromagnetic environment. The exercise represents a significant escalation from Silent Swarm 2025, which was primarily a US Navy demonstration. The 2026 iteration will bring together allied forces to test layered Short-Range Air Defense, SHORAD, integrated with cognitive electronic warfare platforms and AI-enabled autonomous defensive systems.

The cognitive EW platforms that NATO planners are integrating for the exercise represent the cutting edge of electronic warfare: software-defined radios capable of real-time signal classification, adaptive jamming that responds to threat behavior rather than following pre-programmed scripts, and the ability to dynamically spoof Global Navigation Satellite System signals or disrupt drone control links based on what the AI observes the incoming swarm doing. These are not the jammers of the previous generation, brute-force systems that flooded a frequency band with noise. These are intelligent systems that analyze the electromagnetic environment, identify the specific signals they need to attack, and adapt their techniques in real time as the threat adapts.

The choice of Estonia is deliberate. NATO's eastern flank faces the most immediate drone swarm threat from Russia, and Estonia, small, flat, and bordering Russian territory, is the kind of environment where the first autonomous swarm engagement in a European conflict is most likely to occur. The exercise is a dress rehearsal for a scenario that military planners across the alliance believe is not hypothetical but imminent: a contested environment where hundreds of autonomous drones, operating without continuous human control, cross a NATO border and must be detected, classified, and defeated by defensive systems that are themselves increasingly autonomous, all within minutes of the first detection.

The exercise sits at the intersection of every development described above. The drones attacking the defended position will use the same technologies driving the Ukrainian arms race, autonomous navigation, cognitive radio, onboard target recognition. The defenses will use the same distributed EW concepts demonstrated by TEEMS at Silent Swarm 2025. The entire engagement will unfold at machine speed, in a contested electromagnetic environment, with AI on both sides of the exchange.

The humans will set the parameters. The machines will fight the fight.

The Hidden Driver

Electronic warfare is the invisible hand that shapes every drone engagement on every battlefield where unmanned systems operate. It determines whether the hunter can see and whether the killer can strike. It is the environmental pressure that drives the selection of autonomous systems over human-controlled ones, the same way that a predator drives the evolution of speed in its prey. The jammer does not intend to make drones autonomous. It does not need to. It simply kills the drones that depend on human operators, and the drones that survive are the ones that do not.

This is the pattern that connects a Northrop Grumman engineer watching a laptop at Silent Swarm 2025 to a Ukrainian FPV operator watching his screen go to static in a basement in Zaporizhzhia. The electromagnetic spectrum is the medium. Autonomy is the adaptation. And the adaptation, once begun, does not reverse. A drone that has been given the capability to complete a strike without its operator will not have that capability removed because it makes ethicists uncomfortable. A swarm that has been given the ability to coordinate its own electronic warfare without human input will not be slowed down to wait for a human decision that arrives too late to matter. The technology moves in one direction.

The drones can now see in the dark, navigate without GPS, coordinate without central control, and fight through the densest jamming environment in the history of electronic warfare. The five chapters of Part II have traced the technology that makes this possible: the hunter drones that find targets, the killer drones that strike them, the mothership platforms that deliver swarms at scale, the persistent surveillance architectures that never stop watching, and the electronic warfare systems that determine which of these capabilities survive contact with the enemy. The technology is real. It is deployed. It is improving at a pace that outstrips the ability of any government, any treaty body, any ethics commission to keep up.

But technology does not exist in a vacuum. The same AI that helps a Skydio X10D navigate a GPS-denied building also helps an algorithm in Tel Aviv flag 37,000 people for potential killing. The same autonomous targeting software that lets a Ukrainian FPV drone complete a strike after its operator is jammed also lets an intelligence system generate kill lists with 20-second human review. The same distributed coordination that makes a TEEMS swarm resilient also makes it impossible for a single human to understand, much less control, what every node in the network is doing at any given moment.

Part III turns to the hardest questions. Not what the technology can do: that story has been told. But what happens when it is used. What happens when the Hunter-Killer gap compresses to zero and the machine acts without waiting for the human. What the evidence from Gaza reveals about AI targeting at scale. What proliferation means when the same $200 drone that defends a democracy can be turned against one. And why the stakes of getting this wrong are measured not in defense budgets or market capitalizations, but in human lives.

PART III: THE STAKES

Chapter 9: The Kill Algorithm

In an Israeli military intelligence office in late October 2023, three weeks after the Hamas-led assault on southern Israel that killed nearly 1,200 people and took roughly 240 hostage, an officer sat at a terminal and reviewed a list of names. The names had not been compiled by human analysts working through intelligence files, cross-referencing sources, deliberating over evidence. They had been generated by an algorithm. The system on his screen was called Lavender, a machine-learning program developed by Unit 8200, Israel's elite signals intelligence unit, that processed mass surveillance data, assigned each Palestinian in Gaza a probabilistic score based on behavioral patterns, and flagged those whose patterns resembled known militants. Lavender had generated a list of 37,000 names.

The officer's job was to review each target before authorizing a strike. He spent 20 seconds per name. In that time, he confirmed one thing: the target was male. Then he approved the strike and moved to the next name. He did dozens per day. "I would invest 20 seconds for each target at this stage, and do dozens of them every day," the officer later told *+972 Magazine* and *Local Call*, the Israeli-Palestinian outlets whose investigative journalist Yuval Abraham broke the story in April 2024. "I had zero added-value as a human, apart from being a stamp of approval. It saved a lot of time."

Twenty seconds. The time it takes to read a short text message. The time it takes to tie a shoe. In those 20 seconds, a human being was supposed to exercise the judgment that international humanitarian law demands before authorizing lethal force, the assessment of whether this specific individual was a lawful target, whether the expected civilian casualties were proportionate to the military advantage, whether all feasible precautions had been taken. The officer did not review the underlying intelligence. He did not examine the data correlations that led the algorithm to flag this name. He did not know, and could not know in 20 seconds, whether the person on his screen was a Hamas operative or a civil defense worker, a militant or a police officer, a fighter or a father whose phone data happened to resemble that of a fighter. He confirmed the gender and moved on. Twenty seconds per name, dozens per day, 37,000 names on the list.

Another officer offered a different explanation for why the system worked the way it did. "The machine did it coldly," he told the reporters. Without the moral hesitation a human might feel. Without compunction. The machine processed patterns. The machine assigned scores. The machine did not pause to wonder whether the person at the end of the data trail was guilty or innocent. It was not designed to wonder. And the humans operating the machine, by their own account, had stopped wondering too.

This is what happens when the Hunter-Killer gap compresses to zero. Not on some future battlefield. Not in a theoretical scenario debated by ethicists at a conference. In Gaza, in late 2023 and 2024, with real bombs dropped on real houses.

The Machine That Generates Targets

To understand what happened in Gaza, it is necessary to understand three systems, each performing a different function in the kill chain, each feeding the next.

The first was the Gospel: known in Hebrew as *Habsora*. Developed by Unit 8200 and publicly acknowledged by the Israeli military, the Gospel is an AI-based decision support system that automatically reviews massive quantities of surveillance data to identify buildings, equipment, and positions associated with the enemy and recommends them as bombing targets. Retired Lieutenant General Aviv Kohavi, who served as head of the Israel Defense Forces until January 2023, described the Gospel as "a machine that produces vast amounts of data more effectively than any human, and translates it into targets for attack." In an interview published before the current war, Kohavi stated that when the Gospel was activated during Israel's 11-day conflict with Hamas in May 2021, it generated 100 targets per day, with 50% of them being attacked. To appreciate the scale of that acceleration: human analysts had previously produced approximately 50 targets in Gaza per year. The Gospel produced that number before breakfast.

The second system was Lavender. While the Gospel recommended buildings and infrastructure, Lavender identified individual human beings. The algorithm processed communications data, travel patterns, associations, and, according to multiple reports, WhatsApp group membership to assign each person a probabilistic score indicating the likelihood that they were a member of Hamas or Palestinian Islamic Jihad. If an individual's data patterns resembled those of known militants, the system flagged them for potential attack. During the initial weeks of the Gaza war, according to the testimony of six Israeli intelligence officers, the IDF relied almost entirely on Lavender for target selection, treating the algorithm's output as a kill list and adopting it with what one source described as "sweeping approval."

The third system bore the name that reveals the most about the logic of the entire enterprise. "Where's Daddy?" was an AI tracking program that monitored the movements of individuals flagged by Lavender and waited until they returned to their family homes (usually at night, when their entire families were present) before triggering a strike. An Israeli official speaking to *+972 Magazine* stated that "the IDF bombed them in homes without hesitation, as a first option. It's much easier to bomb a family's home. The system is built to look for them in these situations."

Three systems. One to identify buildings for destruction. One to identify people for killing. One to track those people to the place where killing them would also kill their families. Together, they constituted the most extensively documented use of artificial intelligence in lethal targeting in the history of warfare.

The Error Rate They Accepted

The IDF knew Lavender was not perfect. During the system's development, Unit 8200 randomly sampled the algorithm's predictions and concluded that Lavender had achieved a 90% accuracy rate. The military approved its sweeping use as a target recommendation tool on the basis of that figure.

A 90% accuracy rate sounds high. It is the kind of number that, in a product demo or an academic paper, suggests a system that works. Applied to a list of 37,000 names, it means something different. It means 3,700 people on the list, one in 10, were false positives. They had loose or no connections to militant groups. They were civilians. And they were marked for killing by an algorithm whose internal logic was opaque even to the officers who used it.

The error rate was not a secret. It was a known parameter, accepted as the cost of speed. One intelligence source, justifying the approach, told *The Guardian*: "When it comes to a junior militant, you don't want to invest manpower and time in it." Another was more explicit: "You're willing to take the margin of error of using artificial intelligence, risking collateral damage and civilians dying, and risking attacking by mistake, and to live with it."

The Israeli government has contested key elements of this reporting. In a formal response to *The Guardian,* the IDF stated that it "does not use an artificial intelligence system that identifies terrorist operatives or tries to predict whether a person is a terrorist," describing the system instead as "simply a database whose purpose is to cross-reference intelligence sources." The IDF stated that "analysts must conduct independent examinations" to verify targets meet legal definitions, and that "the IDF directs its strikes only towards military targets and military operatives and carries out strikes in accordance with the rules of proportionality and precautions in attacks."

The testimony of six intelligence officers who operated the system tells a different story. And the civilian death toll in Gaza, which by early 2025 exceeded 40,000 according to Gaza health authorities, and continued to rise, with the majority women and children, is consistent with a targeting methodology that accepted a 10% error rate across tens of thousands of targets, permitted 15 to 20 civilian deaths per strike on a low-ranking operative, and deliberately targeted family homes with unguided munitions.

The Twenty-Second Stamp

The most consequential detail in the *+972 Magazine* investigation is not the algorithm itself. Algorithms can be refined, retrained, constrained. The most consequential detail is what happened, or rather, what did not happen, between the algorithm's recommendation and the bomb's release.

In previous Israeli military operations, the process of identifying a human being as a legitimate target involved deliberation. Multiple sources described a system where the decision to "incriminate" an individual (the IDF's internal term for classifying someone as a lawful target) would be discussed among intelligence analysts and then signed off by a legal adviser. The process was slow, and it was designed to be slow. The slowness was the safeguard. It created space for doubt, for questions, for the recognition that intelligence is uncertain and that the wrong answer means a dead civilian.

After October 7, that process was dismantled. Commanders demanded a continuous pipeline of targets. "We were constantly being pressured: 'Bring us more targets.' They really shouted at us," one intelligence officer told *The Guardian*. "We were told: now we have to fuck up Hamas, no matter what the cost. Whatever you can, you bomb."

To meet that demand, the IDF deferred to Lavender. And the human review that was supposed to provide the legal and moral check on the algorithm's output collapsed into the 20-second confirmation described by the officer at the beginning of this chapter. The review was not a review. It was a verification that the name on the screen corresponded to a male. Nothing more.

This is the phenomenon that researchers call **automation bias**: the well-documented tendency for humans to trust computerized decisions and downplay the possibility of machine error. The phenomenon is not new to military operations. In 2003, during the opening days of the Iraq invasion, US Army Patriot missile batteries shot down two allied aircraft, a British Tornado and a US Navy F/A-18: killing three servicemembers. Investigations determined that false confidence in the Patriot system's friend-or-foe identification led operators to trust the machine's classification over their own judgment. The machine said hostile. The operators fired. The targets were friendly. Automation bias, applied to air defense, killed allied pilots. Applied to a targeting system that processes 37,000 names, the consequences scale proportionally.

Dr. Marta Bo, a researcher at the Stockholm International Peace Research Institute, warned that even when "humans are in the loop," operators risk developing automation bias and "over-rely on systems which come to have too much influence over complex human decisions." Richard Moyes of Article 36, a group that campaigns to reduce harm from weapons, stated that when relying on AI targeting tools, a commander "is handed a list of targets a computer has generated" and "doesn't necessarily know how the list has been created or have the ability to adequately interrogate and question the targeting recommendations." He added: "There is a danger that as humans come to rely on these systems they become cogs in a mechanised process and lose the ability to consider the risk of civilian harm in a meaningful way."

Tal Mimran, a lecturer at Hebrew University in Jerusalem who has previously worked with the Israeli government on targeting, noted that the sheer volume of targets generated by AI systems puts pressure on human reviewers that makes automation bias almost inevitable. Analysts "may be tempted to make life easier for themselves by going along with the machine's recommendations," Mimran warned, which could create "a whole new level of problems" if the machine is systematically misidentifying targets.

The danger these researchers described is not theoretical. In Gaza, it was operational reality. The human was in the loop. The human spent 20 seconds in the loop. The human had zero added value.

Dumb Bombs for Unimportant People

The targeting methodology was paired with a munitions policy that compounded its effects.

When the IDF struck the junior operatives identified by Lavender, the thousands of low-ranking individuals who would never have appeared on traditional target lists, the military typically used unguided munitions. "Dumb bombs." Free-fall explosives dropped from aircraft, guided only by onboard release-point calculations, inherently less precise than the guided munitions the Israeli Air Force is capable of deploying. The reason was economic. "You don't want to waste expensive bombs on unimportant people, it's very expensive for the country and there's a shortage [of those bombs]," one intelligence officer explained.

The Israeli Air Force possesses some of the most precise guided munitions in any military's arsenal. It can strike a single room in a multistory building. It chose not to use that precision on the targets Lavender identified as junior operatives, the vast majority of the 37,000 names on the list. Instead, the military dropped unguided bombs on residential apartment buildings, weapons whose blast radius and fragmentation pattern make it physically impossible to destroy only the intended target. The consequences are predictable and were predicted. An unguided bomb dropped on an apartment building destroys the building and kills everyone inside. The officer acknowledged as much: "We usually carried out the attacks with dumb bombs, and that meant literally dropping the whole house on its occupants."

The pre-authorized collateral damage ratios confirmed the calculus. During the early weeks of the war, according to *+972 Magazine* and *The Guardian*, the IDF permitted 15 to 20 civilian deaths per airstrike aimed at a single low-ranking militant. For senior Hamas commanders, the acceptable threshold rose to over 100 civilian casualties per strike. One source described a formal hierarchy: "We had a calculation for how many [civilians could be killed] for the brigade commander, how many for a battalion commander, and so on."

The combination was systematic. An algorithm with a 10% error rate generated a list of 37,000 names. Human reviewers spent 20 seconds per name, confirming only gender. The military tracked targets to their family homes and bombed them at night with unguided munitions, pre-authorizing the deaths of 15 to 20 civilians per strike. One intelligence source captured the operational logic in a single sentence: "Because of the system, the targets never end. You have another 36,000 waiting."

The Prophecy of the Machine Team

The architecture of AI-driven targeting in Gaza did not appear from nowhere. Its intellectual foundation was published three years before the war began.

In 2021, a book titled *The Human-Machine Team: How to Create Synergy Between Human and Artificial Intelligence That Will Revolutionize Our World* appeared under the pen name "Brigadier General Y.S." The author was later identified as Yossi Sariel, the commander of Unit 8200, the same unit that developed Lavender and the Gospel. In the book, Sariel described a vision in which AI systems could generate 80,000 targets before combat and 1,500 targets per day during an active war. He advocated for a model in which human intelligence analysts would be augmented, and largely replaced, in terms of volume, by machine-learning algorithms capable of processing surveillance data at a scale no team of humans could match.

The book was published before the current war. Its claims seemed ambitious at the time. In retrospect, they read as a blueprint. The Gospel was producing 100 targets per day during the 2021 conflict. Lavender generated 37,000 individual targets in the opening weeks of the 2023 war. During the first 35 days of fighting after October 7, the Israeli Air Force attacked some 15,000 targets, a rate of strikes exceptionally higher than any previous Israeli military campaign. The trajectory Sariel described was not speculative. It was the operational plan of the intelligence unit he commanded, executed at a scale he had publicly predicted.

Sariel's identity was later exposed through an online security lapse reported by *The Guardian* in April 2024, the same month the Lavender investigation was published. The coincidence of timing (the architect's unmasking alongside the revelation of what his architecture had produced) illuminated a connection that runs through this entire book. The people building these systems are not hiding their intentions. They are publishing them, in English, under pseudonyms that barely conceal their identities, describing capabilities that match almost exactly what was later deployed. The question is whether anyone reads the blueprint before the building is occupied.

The Accountability Gap

International humanitarian law rests on a simple premise: human beings are responsible for their decisions in war. A soldier who commits a war crime can be prosecuted. A commander who orders an unlawful strike can be held liable. The chain of responsibility runs from the trigger to the top, and every link in that chain is a person who made a choice.

AI targeting systems fracture that chain.

When Lavender flags a target, who is responsible for the decision? The algorithm cannot be prosecuted. The engineers who designed the system did not select the specific target. The officer who spent 20 seconds confirming the target was male did not understand the algorithm's logic, did not know what data correlations it drew on, what weight it assigned to various inputs, why this name appeared on the list instead of another. The commander who approved the system's use established the policy but did not review individual targets. The intelligence chief who built the system wrote a book advocating for its creation but was not in the room when the bombs fell.

Human Rights Watch and other organizations have identified this as the **accountability gap**, the risk that when militaries use autonomous or semi-autonomous decision support systems, no person can be meaningfully held liable for the resulting harm. The internal logic of the AI is opaque. Its data correlations are inscrutable. Prosecutors and judges will struggle to prove which decisions were manifestly unreasonable, or who held the requisite intent or knowledge for criminal liability. The machine's opacity becomes a shield.

This is not a flaw in the system. From a certain institutional perspective, it is a feature. If no single person made the decision, no single person bears the blame. The algorithm distributes responsibility so widely that it becomes effectively invisible. A commander who personally reviewed intelligence files and signed off on a strike that killed civilians can be held accountable for that specific judgment. A commander who activated Lavender and told his officers to spend 20 seconds per name can point to the algorithm, the error rate, the operational pressures, the policy guidance he received from above. The machine did not remove human judgment. It created a structure in which human judgment was no longer necessary, and therefore no longer accountable.

One intelligence officer, asked by *+972 Magazine* about the moral dimension of trusting the algorithm, offered a response that belongs in every textbook on automation bias for the next generation: "The machine did it coldly."

The Other Side of the Argument

This chapter has presented evidence that is devastating. A serious treatment of AI targeting in Gaza requires engaging with the strongest version of the Israeli position, not the weakest. That position is not frivolous.

The October 7 attack was the deadliest day for Jewish people since the Holocaust. Hamas militants killed nearly 1,200 Israelis, many of them civilians at a music festival and in their homes. Approximately 240 people were taken hostage. The Israeli government and the IDF have described the subsequent military campaign as a response to an existential attack carried out by a terrorist organization that embeds its military operations within civilian infrastructure, command posts in schools, tunnels under hospitals, rocket launchers in residential neighborhoods. In this context, the argument goes, the speed and scale of AI targeting was not reckless but necessary. Hamas operates in one of the most densely populated places on Earth, making the separation of combatants from civilians extraordinarily difficult. The alternative to algorithmic targeting at speed, in this view, is slower targeting that allows Hamas to relocate, rearm, and launch further attacks. Lavender, from this perspective, did not replace human judgment so much as augment human capacity at a moment when the volume of threats exceeded what any team of analysts could process.

The IDF has also argued that the systems described in the *+972 Magazine* investigation are not what the reporting claims. The military's formal response stated that the system is "not a list of confirmed military operatives eligible to attack" but rather a database for cross-referencing intelligence sources. The IDF stated that analysts must conduct independent examinations and that the military uses munitions "while employing onboard aircraft systems to calculate a specific release point to ensure a high level of precision." The IDF maintains that its operations comply with international humanitarian law.

These arguments deserve to be heard. They represent the position of a democratic government operating under extraordinary duress after a catastrophic attack on its civilian population. The IDF is not wrong that Hamas embeds its military operations within civilian infrastructure, or that this complicates the application of international humanitarian law.

And the arguments do not survive contact with the evidence.

International humanitarian law does not contain an exception for urgency. The principles of distinction and proportionality apply regardless of the scale of the threat, regardless of the density of the combat environment, regardless of the pressure commanders feel to generate targets. The obligation to distinguish between combatants and civilians, codified in Article 48 of Additional Protocol I and reflected in customary international law, requires that each individual attack be assessed on its own terms. A system that generates 37,000 targets with a known 10% error rate, reviewed for 20 seconds each, struck with unguided bombs in family homes at night, with pre-authorized civilian casualty ratios of 15 to 20 per low-ranking militant, does not meet that standard. It does not come close.

The question is not whether the IDF faced genuine operational challenges. It did. The question is whether AI targeting systems as deployed in Gaza met the legal and moral standards that every military claiming to comply with the laws of war is obligated to uphold. The evidence, as documented by *+972 Magazine*, corroborated by *The Guardian*, and assessed by Human Rights Watch, the International Committee of the Red

Cross, and legal scholars, indicates that they did not.

Human Rights Watch has concluded that autonomous and semi-autonomous weapons systems violate the right to life by arbitrarily depriving individuals of it, violate human dignity by reducing people to data points in an algorithm, and create what the organization calls "digital dehumanization", the process by which human beings are stripped of their identity and worth, reduced to coordinates and probability scores, and subjected to lethal force by a system that cannot comprehend the value of what it destroys. The ICRC has called for new legally binding rules to regulate autonomous weapon systems, and together with UN Secretary-General Antonio Guterres has urged states to conclude treaty negotiations by 2026.

A note on sourcing is warranted here. The core reporting on Lavender comes from *+972 Magazine* and *Local Call*, based on the testimony of six anonymous Israeli intelligence officers. Anonymous sourcing always warrants caution. But the reporting has been independently corroborated by *The Guardian*, which conducted its own verification of the sources and the claims. Human Rights Watch, academic researchers at multiple institutions, and the IDF's own public statements (which acknowledge the existence of the systems while contesting their characterization) all support the essential architecture of the account. The IDF's response does not deny the existence of the systems. It disputes what they are called and how they are described. That is a significant distinction.

The author of this book takes a position. The evidence from Gaza demonstrates that AI targeting as deployed in the 2023-2024 war compressed the Hunter-Killer gap to the point where meaningful human judgment was effectively eliminated.

The algorithm identified. The algorithm tracked. The human rubber-stamped. The bomb fell. The system operated not as a tool augmenting human decision-making but as a pipeline that replaced it, a pipeline in which the human was retained not for judgment but for legal cover, a 20-second formality in a kill chain that ran from sensor to algorithm to bomb with no substantive human intervention. That is not a human-on-the-loop system. That is an autonomous targeting system with a human-shaped stamp on the paperwork.

The Anthropic Test

If the Gaza evidence represents the most consequential documented case of AI targeting already deployed, a confrontation in Washington in early 2026 revealed how the tension between AI capability and AI restraint is playing out inside the country that builds the most powerful models on Earth.

In January 2026, Secretary of Defense Pete Hegseth issued an AI strategy memo instructing that every Defense Department AI contract must include standard "any lawful use" language, effectively requiring the removal of any vendor-imposed safety restrictions beyond what law requires. The memo stated that "the risks of not moving fast enough outweigh the risks of imperfect alignment."

The directive collided with Anthropic, the AI safety company whose model, Claude, had been awarded a $200 million prototype contract in July 2025 and was deployed on classified military networks through Palantir's Maven Smart System. On February 26, 2026, Anthropic CEO Dario Amodei laid out two absolute red lines: Claude could not be used for mass domestic surveillance of US citizens, and it could not power fully autonomous weapons operating without human oversight. "Frontier AI systems are simply not reliable enough to power fully autonomous weapons," Amodei stated. The oversight mechanisms needed to protect civilian and military lives "don't exist today."

The Pentagon refused. On February 27, Hegseth directed the Defense Department to designate Anthropic as a "supply chain risk to national security." President Trump directed every federal agency to cease all use of Anthropic's technology. Pentagon spokeswoman Kingsley Wilson stated that US forces would not be "held hostage by Silicon Valley ideology." Hours later, OpenAI announced a deal with the Pentagon accepting the any-lawful-use baseline, though OpenAI CEO Sam Altman later acknowledged the agreement was "rushed" and "looked opportunistic and sloppy." OpenAI subsequently amended the contract after public and employee backlash.

The episode illuminated a paradox at the center of the autonomous weapons debate. Anthropic built the safety-first AI company. It drew red lines on autonomous weapons and domestic surveillance. And it was blacklisted for it, while the military reportedly continued using Claude in active operations because commanders had become too dependent on the system to give it up. The company that said "not reliable enough" was punished. The company that said "any lawful use" was rewarded. The market incentive and the safety incentive pointed in opposite directions.

The paradox deepened further. Amodei's red lines were narrow. He explicitly stated that Anthropic supported the use of Claude for "lawful foreign intelligence and counterintelligence", meaning the same model that could not be used for domestic surveillance could, under Anthropic's own policy, assist with foreign surveillance and targeting assistance. The company's ethics stopped at the US border. But even those limited restrictions were too much for the Pentagon. The message was clear: in the current procurement environment, any safety guardrail is one guardrail too many.

Anthropic has since filed suit in federal court, challenging its designation as a supply chain risk. The case raises questions that extend far beyond a single contract. If the only AI companies the Pentagon will work with are the ones willing to remove every safety guardrail, then the guardrails that might prevent a Lavender-scale deployment by the United States will be stripped away not by a policy decision but by a procurement decision, not debated in Congress but negotiated in a contract vehicle that most Americans will never see.

The Black Box and the Bomb

The thread connecting an intelligence officer in Tel Aviv spending 20 seconds on a name to a Pentagon procurement memo demanding "any lawful use" is not a coincidence. It is a pattern. And the pattern has a direction.

The technology that enables AI targeting is improving. The models are getting more capable. The data pipelines are getting faster. The integration between surveillance systems and strike platforms is getting tighter. Every element of the kill chain that once required human time (the time to analyze, the time to deliberate, the time to doubt) is being compressed by software designed to make the chain faster.

The Gaza evidence reveals what happens at the end of that compression. The human does not disappear from the process. The human remains, technically, in the loop. But the loop has been shrunk to 20 seconds, and in those 20 seconds the human performs no function that a simple gender-verification algorithm could not perform faster and more consistently. The human is not exercising judgment. The human is providing the appearance of judgment, the legal and institutional cover that allows the system to be described as "human in the loop" while functioning, in practice, as if the human were not there at all.

This is the lesson that every military planning to deploy AI targeting systems must absorb. The technology does not eliminate the human by removing the human from the process. It eliminates the human by making the human irrelevant within the process, by generating targets so fast, in such volume, with such apparent confidence, that the human operator becomes a rubber stamp, a cog, a 20-second delay between the algorithm and the bomb. The label "human in the loop" becomes a legal fiction, technically accurate, operationally meaningless.

The International Committee of the Red Cross, Human Rights Watch, and the Campaign to Stop Killer Robots have all called for a legally binding international treaty to ensure meaningful human control over the use of force. The word "meaningful" is doing all the work in that sentence. A 20-second review is human control. It is not meaningful human control. The distinction between those two concepts (between the presence of a human and the exercise of human judgment) is the most important line in the autonomous weapons debate. Every military that claims its AI systems keep a human in the loop must be asked: what does the human do? How long does the human spend? What information does the human have? What authority does the human exercise? If the answers are "confirms gender," "20 seconds," "whatever the algorithm shows," and "none that the algorithm could not exercise faster," then the human is not in the loop. The human is decoration.

The UN General Assembly voted 166 in favor, 3 against, with 15 abstentions in December 2024 on a resolution on autonomous weapons. But resolutions are not treaties, and the nations developing the most advanced AI targeting systems, the United States, Israel, Russia, China, have shown no willingness to accept binding constraints. The Convention on Certain Conventional Weapons, the forum where autonomous weapons have been debated since 2014, operates by consensus, meaning a single nation can block progress. The Group of Governmental Experts on lethal autonomous weapons systems is approaching the end of its three-year mandate, with the CCW Seventh Review Conference in November 2026 as the decisive deadline. Progress has been glacial. The technology has not waited.

Israel is not the only country building these systems. It is not even the most advanced. The same machine-learning techniques that power Lavender (pattern recognition applied to surveillance data, probabilistic scoring, automated classification) are available to any government or defense contractor with sufficient data and computational resources. The United States, China, Russia, the United Kingdom, France, South Korea, and Turkiye are all developing or deploying AI-enabled targeting systems in various configurations. The same automation bias that collapsed human review in Gaza into a 20-second rubber stamp will operate identically in any military that trusts an algorithm to generate targets faster than humans can evaluate them. The psychology is universal. The institutional pressure for speed is universal. The same accountability gap that shields commanders from prosecution when the algorithm's logic is opaque will shield commanders in Washington, Moscow, Beijing, and Ankara with equal effectiveness. The black box does not recognize borders.

The technology that produced Lavender, Gospel, and "Where's Daddy?" is not an Israeli technology. It is a universal technology. And the question it poses is not about one country's conduct in one war. It is about what happens when every country has the capability to generate kill lists at machine speed, with 20-second human review, and no one can explain how the algorithm chose the names.

That question is about to become much harder to answer. Because the same capabilities that enabled AI targeting in Gaza are proliferating, not just to governments with billion-dollar intelligence budgets, but to anyone with a $200 drone and the software to point it at a target. The next chapter examines what happens when the kill algorithm goes global.

PART III: THE STAKES

Chapter 10: The Democratization of Lethality

The workbench was a folding table from a hardware store, set up in a garage in a city that does not need to be named. The components spread across it, in the spring of 2025, could have been ordered from any e-commerce platform on earth. A DJI flight controller, removed from its original housing and wired into a custom frame. A set of first-person-view goggles, the same model used by racing drone hobbyists in forty countries. Four lithium-polymer battery cells, shrink-wrapped and labeled in Mandarin. A 3D printer humming in the corner, extruding a stabilizer fin from a CAD file downloaded off a public repository. And at the center of the table, the one component that separated this project from a weekend hobby build: a repurposed warhead, sourced from a conventional mortar round, fitted with a simple impact fuze.

Total cost of the components on the table: under $500. Total assembly time for a person with moderate electronics experience: four to six hours. The device being assembled was not a toy, not a racing drone, not a delivery vehicle. It was a precision-guided weapon capable of destroying an armored vehicle, penetrating a hardened fighting position, or killing a specific individual from several kilometers away. No export license was required for any component except the warhead. No government approved the sale of the flight controller or the goggles or the batteries. No international treaty covered the transaction. No arms dealer was involved.

This is what the democratization of lethality looks like. Not a clandestine weapons factory. Not a state-sponsored laboratory. A folding table, a 3D printer, and an internet connection. The components of a precision strike weapon, a capability that fifteen years ago required a Predator drone, a satellite link, a ground control station in Nevada, and a military budget measured in billions, are now available on the open market for the cost of a midrange smartphone. The Hunter-Killer gap, the distance between finding a target and striking it, is being compressed not only by great powers running billion-dollar autonomy programs. It is being compressed by individuals with soldering irons and open-source software.

The previous chapters traced how that gap is closing at the technological frontier, through AI-driven autonomy, mesh-networked swarms, and electronic warfare adaptations that push drones toward independence from their operators. This chapter examines a different and in some ways more dangerous dimension of the same revolution: the proliferation of precision strike capability to actors who were never supposed to have it. States that could not afford a modern air force now field drone fleets capable of destroying armor columns. Non-state groups that once relied on improvised explosive devices and small arms now strike targets with guided munitions. And individuals with no affiliation to any military, operating from garages and workshops, can assemble weapons that perform the same function as systems costing thousands of times more.

The democratization of lethality is proceeding along three vectors, state proliferation, non-state adoption, and do-it-yourself production, and no existing arms control framework is equipped to stop any of them.

The Export Champions

The first vector is the oldest and the most familiar: one state selling weapons to another. What is new is the speed, the scale, and the identity of the sellers.

For decades, the global arms market for advanced military aircraft was dominated by a small club of exporters: the United States, Russia, France, the United Kingdom, and, increasingly, Israel. The barriers to entry were enormous. Developing a combat aircraft required billions in research and development, decades of engineering expertise, and a domestic military willing to serve as the first customer. The export market was tightly controlled. The United States restricted sales of its most capable drones under the Missile Technology Control Regime (MTCR), a 1987 agreement originally designed to prevent the spread of nuclear-capable ballistic missiles. The MTCR placed its strictest controls on systems capable of carrying a 500-kilogram payload over 300 kilometers, thresholds that captured the MQ-9 Reaper and effectively prevented Washington from selling armed drones to most of its allies for years.

Then Turkiye's Baykar entered the market, and the old order cracked.

The Bayraktar TB2, a medium-altitude long-endurance unmanned combat aerial vehicle developed largely by Selcuk Bayraktar, a former MIT graduate student, became the most exported combat drone in the world. By 2025, the TB2 had been sold to the militaries of at least 34 countries and surpassed one million flight hours globally. Baykar recorded $2.2 billion in export revenue. The unit cost: roughly $5 million to $5.5 million per aircraft, about one-sixth the price of a Reaper, made armed drone capability accessible to nations that had never possessed anything comparable. Poland became the first NATO and European Union member to purchase the TB2, ordering 24 in 2021. Romania bought three systems totaling 18 drones for $321 million. Nigeria, Pakistan, Morocco, Rwanda, Togo, Niger, Burkina Faso, Mali, Kenya, Bangladesh, Kuwait, the United Arab Emirates, the customer list read like a map of the developing world's security anxieties, and Baykar filled every order.

The TB2 did not just sell well. It fought well, and the combat footage traveled faster than any export brochure. During the 2020 Nagorno-Karabakh war, Azerbaijan used TB2s to systematically dismantle Armenian defenses, destroying artillery, BM-30 Smerch rocket launchers, T-72 tanks, infantry fighting vehicles, and nine Osa and Strela-10 air defense systems. The footage, thermal imagery of armored vehicles detonating, broadcast on Azerbaijani social media in near-real time, became the most consequential drone warfare advertising campaign in history. In Libya, the UN-recognized Government of National Accord used TB2s to neutralize Russian-made Pantsir air defense systems, destroying at least nine units. In Ethiopia, TB2s were deployed against the Tigray People's Liberation Front; in January 2022, a drone strike on a camp for internally displaced people in Dedebit killed nearly 60 civilians. Investigators recovered debris from a MAM-L guided munition at the site, a weapon exclusively used with the TB2.

Each conflict expanded the market. Each combat demonstration proved the capability. Each sale put precision strike in the hands of another military that, a decade earlier, could not have imagined possessing it.

Baykar was not alone. China filled the market space that American export restrictions had created. The China Aerospace Science and Technology Corporation (CASC) produced the CH-series drones, the CH-4 and CH-5, medium-altitude long-endurance platforms comparable to the Reaper in concept if not in sophistication, and sold them to militaries across Africa, the Middle East, and Central Asia. The Aviation Industry Corporation of China, AVIC, manufactured the Wing Loong series for similar markets. In 2019, US Defense Secretary Mark Esper publicly criticized China for exporting drones capable of autonomously taking human life without human oversight to the Middle East, a charge that highlighted both the proliferation problem and America's inability to stop it. Beijing's export controls were minimal. Its customer vetting was nonexistent by Western standards. And its prices undercut every competitor.

The math was straightforward: countries that could not buy American drones bought Turkish or Chinese ones instead. The MTCR, designed to prevent the spread of nuclear-capable delivery systems, had no mechanism to address a $5 million armed drone that fell well below its payload and range thresholds. The weapon had outrun the regulation.

Tehran's Assembly Line

The second state proliferator operated on a different model entirely. Iran did not sell its drones on the open market to allied governments through formal defense contracts. It distributed them to proxies, partners, and clients as instruments of strategic influence, and in doing so, demonstrated that a mid-tier military power could reshape regional security through drone exports costing a fraction of what conventional arms transfers required.

The centerpiece of Iran's drone proliferation was the HESA Shahed 136, a one-way attack drone, a loitering munition designed to fly to a set of coordinates and detonate on impact. The Shahed was not sophisticated. It used a small piston engine, basic inertial navigation supplemented by GPS, and a warhead estimated at 40 to 50 kilograms. It was slow, loud, and easy to detect. It was also cheap, simple to manufacture in volume, and effective enough to reshape the economics of two separate conflicts.

Russia became Iran's most consequential drone customer. Beginning in the summer of 2022, Iran supplied Shahed 136 drones to Russia for use against Ukraine. Russia designated them "Geran-2" and launched them in waves against Ukrainian power infrastructure, civilian buildings, and military targets. The drones were not precision weapons in the way an American JDAM or a Bayraktar-launched MAM-L was a precision weapon. They were area-effect terror instruments, cheap enough to launch in salvos of dozens, forcing Ukraine to expend air defense interceptors that cost orders of magnitude more than the drones they destroyed. A single Shahed reportedly cost Iran between $20,000 and $50,000 to produce. A single NASAMS interceptor missile used to shoot one down cost roughly $500,000. The exchange ratio was devastating for the defender. By 2025, Iran continued to supply Russia with armed UAVs, and Russia was simultaneously targeting Ukrainian infrastructure with a combination of one-way attack drones and long-range missiles to degrade the country's national will to resist. In exchange, Iran sought advanced Russian military technology, including Su-35 multirole fighters, using drone exports as a currency for upgrading its own conventional forces.

The Shahed proliferation did not stop at Russia. Iran's network of proxy and partner relationships, what Tehran calls the "Axis of Resistance", provided ready-made distribution channels for drone technology. The Houthi forces in Yemen received Iranian-supplied and Iranian-designed drones including the Qasef-1, Qasef-2K, Samad-2, and Samad-3 series. These were not identical to the Shahed 136, but they shared a common design philosophy: low cost, expendable, effective against targets that lacked the air defenses to stop them.

Hezbollah in Lebanon received similar support. Before Israeli military strikes destroyed at least 50% of Hezbollah's military infrastructure beginning in September 2024, the organization had built a significant drone capability with Iranian assistance. After the strikes, Iran signaled its intention to rearm Hezbollah and continue providing lethal aid to impose costs on Israel and threaten US interests.

The Iranian model demonstrated something that traditional arms control theory had not accounted for: a state could become a significant drone proliferator without possessing a world-class defense industry. Iran's drones were not technologically advanced. They did not need to be. They needed to be cheap enough to produce in quantity, simple enough to transfer to groups with limited technical capacity, and effective enough to impose costs on adversaries whose defense systems cost multiples of the attacking weapon. The Shahed met all three criteria. Tehran had found a way to project power across the Middle East and into the European theater for a fraction of the cost of a single advanced fighter aircraft.

The Houthi Proof of Concept

The most dramatic demonstration of what drone proliferation means for non-state actors came not from a battlefield but from a shipping lane.

Since November 2023, the Yemen-based Houthi forces conducted nearly 200 maritime attacks across the Red Sea, Gulf of Aden, and Indian Ocean. They struck more than 30 merchant vessels, including at least 10 oil tankers. They attempted attacks against US naval ships and caused light damage in Israel. The campaign, executed with a combination of drones, anti-ship missiles, and waterborne explosive devices, decreased Red Sea transit by 70%. Global shipping companies rerouted vessels around the Cape of Good Hope, adding weeks and millions of dollars in fuel costs to every transit between Asia and Europe.

The Houthis were not a state. They controlled a portion of Yemen. Their military budget was opaque but by any measure a rounding error compared to the US Navy fleet that deployed to the region to protect commercial shipping. And yet a non-state actor with Iranian-supplied drones and missiles had imposed a functional blockade on one of the world's most critical maritime chokepoints.

The cost asymmetry was the story. The United States deployed a carrier strike group, guided missile destroyers, and cruisers to the Red Sea. The cost of a single SM-2 Standard missile fired from a destroyer to intercept a Houthi drone is roughly $2 million. The cost of the drone it intercepted was a tiny fraction of that figure. Every successful intercept was a tactical victory and an economic defeat. The Houthis could sustain the campaign indefinitely as long as Iran continued supplying weapons and components. The US Navy could not sustain its interceptor stockpile indefinitely at that exchange rate.

The Houthis used their campaign for multiple purposes simultaneously: to extort shipping companies for protection from strikes, to press for international concessions including ceasefires in Gaza and Lebanon, and to garner domestic political support by projecting military power against the world's preeminent naval force. The campaign succeeded on almost every dimension that mattered to the Houthis. It failed on almost every dimension that mattered to the global economy.

This was the non-state proliferation nightmare made real. A group with no air force, no navy in any conventional sense, no defense industry, and no seat at the United Nations had acquired the capability to disrupt global trade, threaten the world's most expensive warships, and impose costs that the defending power could not sustain at scale. The drones that made it possible cost less than the ammunition used to destroy them.

Five Hundred Garages

The third vector of proliferation is the most novel and the hardest to control. It does not involve state-to-state arms sales or proxy networks. It involves individuals and small groups building precision weapons from commercially available components.

Ukraine demonstrated the model at scale. By 2025, more than 500 separate Ukrainian drone producers were manufacturing combat drones, aerial, ground, and maritime platforms spanning the full spectrum from disposable FPV munitions to long-range strike systems. In 2024, Ukraine produced over 4 million drones. The defense industrial base had built capacity to produce up to 10 million in 2025, with projections reaching 7 million drones annually by 2026.

Mykhailo Fedorov, Ukraine's Minister of Digital Transformation, was the primary government architect behind this decentralized ecosystem. Fedorov spearheaded the "Army of Drones" initiative, which rewrote state procurement rules to slash bureaucratic barriers, lift profit margin caps, and allow hundreds of private companies to sell directly to the military. The government launched Brave1, a defense-tech accelerator platform that functioned as connective tissue between the 500-plus private developers and the armed forces, providing testing grounds, state grants, and direct access to military feedback so that successful prototypes could transition rapidly from garage workbench to battlefield deployment.

The model was not centralized manufacturing. It was an ecosystem, hundreds of independent workshops, some staffed by professional engineers, some by hobbyists and volunteers, some by teenagers who had grown up flying racing drones and now applied that skill to building weapons. A typical garage-scale producer might employ five to fifteen people, assemble 50 to 200 FPV drones per week from commercially sourced components, and ship them to frontline units within days. The production cycle from prototype to deployment was measured in weeks, not the years that traditional defense procurement required. If a design failed in the field, the feedback loop to the producer was days, a soldier's WhatsApp message describing what went wrong, followed by a revised design in the next batch.

The implications extended far beyond Ukraine's borders. The Ukrainian model proved that precision strike capability no longer required an industrial base. It did not require a Lockheed

Martin or a Raytheon or even a Baykar. It required access to commercial components (flight controllers, motors, batteries, cameras, GPS modules, 3D printers) and the engineering knowledge to assemble them into a weapon. That knowledge was not classified. It was not restricted. Much of it was available on YouTube, on Telegram channels, in open-source software repositories, and in the growing body of combat footage that documented what worked and what did not.

The Auterion/Nemyx demonstration in January 2026 illustrated how far the capability had traveled. In what the company described as the world's first live-fire combat drone swarm, a single human operator engaged three separate targets simultaneously with drones built by different manufacturers, coordinated through a unified software interface. The demonstration proved that swarm capability was becoming platform-agnostic, that the software layer could turn any collection of compatible drones into a coordinated strike force, regardless of who manufactured the hardware.

That demonstration was conducted by a professional defense company under controlled conditions. The underlying principle (that software could coordinate heterogeneous drone platforms into a unified kill chain) was equally available to any group capable of writing or acquiring the code.

The Arms Control Void

Every previous revolution in military technology eventually produced an arms control response. Nuclear weapons produced the Non-Proliferation Treaty. Chemical weapons produced the Chemical Weapons Convention. Landmines produced the Ottawa Treaty. Cluster munitions produced the Convention on Cluster

Munitions. The pattern was consistent: a new weapon emerged, its consequences became apparent, and the international community, slowly, imperfectly, but eventually, built a legal framework to constrain it.

The drone revolution has broken this pattern, and the reasons are structural.

The primary international forum for addressing autonomous weapons has been the Convention on Certain Conventional Weapons (CCW), where a Group of Governmental Experts has debated the regulation of lethal autonomous weapons systems (LAWS) for over a decade. The CCW requires consensus to make decisions. That consensus rule has allowed major military powers (the United States, Russia, India, and Israel among them) to persistently block proposals and prevent the adoption of legally binding rules. After more than ten years of negotiations, the CCW has produced no regulatory treaty on autonomous weapons. In September 2024, 42 states delivered a joint statement affirming their readiness to move forward. The path forward remained blocked by the same holdouts that had blocked it for a decade.

In December 2023, the UN General Assembly adopted Resolution 78/241, its first resolution on lethal autonomous weapons systems, with 152 states voting in favor and four, Belarus, India, Mali, and Russia, voting against. In December 2024, 166 nations voted in favor of autonomous weapons regulation. The resolutions mandated a process. They did not produce a treaty. The deadline for concluding negotiations was set for 2026, and no serious observer expected a binding agreement by that date.

The structural problem went deeper than political obstruction. Traditional arms control frameworks were designed for a specific category of weapon: identifiable military hardware, produced by states, transferred through traceable channels, distinguishable from civilian technology. Nuclear warheads. Chemical precursors. Anti-personnel landmines. Each of these weapons had characteristics that made control theoretically possible, even if enforcement was difficult. They were produced in specialized facilities. They required materials or precursors that could be tracked. They were physically distinguishable from civilian equivalents.

Drone technology violates every one of these assumptions. The flight controller in a combat FPV drone is identical to the flight controller in a racing drone. The lithium-polymer batteries that power a weapon are the same batteries that power a hobbyist's camera platform. The 3D-printed components that form a warhead housing are produced on the same printers that produce consumer goods. The computer vision algorithms that enable autonomous targeting are derived from the same machine learning frameworks that power self-driving cars and smartphone cameras. The hardware and software driving lethal drone capability are fundamentally dual-use, the same technology offers both critical civilian and military applications, and no inspection regime can distinguish between them at the component level.

The Missile Technology Control Regime, designed in 1987 to prevent the proliferation of nuclear-capable ballistic missiles, places its strictest controls on systems capable of carrying a 500-kilogram payload over 300 kilometers. Most tactical drones, most FPV munitions, and most loitering munitions fall far below these thresholds. The MTCR was built for an era of exquisite, high-end weapons platforms. It has no vocabulary for a $200 weapon assembled from consumer electronics. The Wassenaar Arrangement, a voluntary export control regime for conventional arms and dual-use goods, relies on the national discretion of its 42 participating states to enforce controls, and its control lists update so slowly that they cannot keep pace with innovation cycles measured in months rather than years.

The result is an arms control void. The international community has spent a decade debating the regulation of weapons that are proliferating faster than any previous weapons category in history, and it has produced no binding constraint on their development, production, transfer, or use. The weapons are too cheap to track, too simple to restrict, too dual-use to distinguish from legitimate commerce, and too useful for too many actors to give up voluntarily.

The Compression from Below

The Hunter-Killer framework, as developed in the preceding chapters, describes how AI compresses the gap between sensing and striking, between finding a target and destroying it. The democratization of lethality adds a second dimension to that compression. The gap is closing not only at the technological frontier but also at the bottom of the capability ladder.

When the United States pioneered precision strike with the Predator drone in the early 2000s, the capability existed at the apex of global military power. Only the United States possessed the combination of satellite communications, GPS-guided munitions, real-time intelligence feeds, and trained operators required to find and kill a specific individual from the other side of the planet. The kill chain was long, expensive, and exclusively

American.

Two decades later, some version of that capability (compressed, simplified, far less precise, but functionally equivalent for many tactical purposes) is available to any state willing to write a check to Baykar or CASC, to any proxy group operating within Iran's distribution network, and to any individual with access to commercial drone components and a basic understanding of electronics. The $200 FPV drone that destroys a $10 million tank is not a Predator. It does not operate from the other side of the planet. It does not have global reach. But it closes the Hunter-Killer gap at the tactical level just as surely as a billion-dollar autonomous system does, and it does so at a price point that makes proliferation unstoppable by any mechanism the international community currently possesses.

The cost ratio tells the story in its starkest form. A Ukrainian FPV drone costing roughly $200 destroyed a Russian Tor air defense system valued at approximately $24 million. The exchange ratio: 1 to 120,000. No arms control regime in history has ever confronted a weapon whose cost-to-effect ratio makes it simultaneously indispensable to the user and impossible to price out of the market. A government can embargo fighter jets. It cannot embargo hobby electronics.

The proliferation of drone technology is the most dangerous dimension of the autonomous weapons revolution because it compounds every other risk. AI targeting becomes more dangerous when the targeting system can be built by anyone. Swarm coordination becomes more threatening when the swarm can be assembled from commercial parts. The ethical questions explored in the previous chapter (about accountability, civilian casualties, and the erosion of human judgment) become orders of magnitude harder when the actors posing those questions are not governments with legal obligations under international humanitarian law but non-state groups with no accountability to anyone.

And the proliferation is accelerating. Every month, the components become cheaper, the assembly knowledge becomes more accessible, and the combat footage from Ukraine and the Red Sea provides a more detailed tutorial in what works. The open-source drone warfare community, a distributed, informal network of engineers, hobbyists, volunteers, and weapons designers operating on Telegram, Signal, and public code repositories, grows faster than any government effort to monitor it.

Traditional arms control assumed that the spread of military capability could be slowed by controlling the supply of weapons and the knowledge to build them. The drone revolution has inverted that assumption. The supply is the global consumer electronics market. The knowledge is the open internet. Neither can be controlled without costs that no government is willing to impose.

What Cannot Be Contained

The uncomfortable conclusion is this: the democratization of lethality is a structural feature of the current technological moment, not a policy failure that better regulation could fix. The same forces that make commercial technology cheap, accessible, and globally distributed, the forces that put a supercomputer in every pocket and a high-definition camera on every drone, also make precision strike capability available to anyone willing to assemble it.

This does not mean that regulation is pointless. Export controls on complete weapons systems still slow state-to-state proliferation. Intelligence operations still disrupt non-state acquisition networks. Norms against autonomous targeting, even unenforceable norms, shape the behavior of states that care about international legitimacy. But it means that the proliferation of drone-enabled precision strike cannot be stopped. It can only be managed, and managed imperfectly, by actors who recognize that the genie is not going back in the bottle.

The proliferation of drone technology is a global phenomenon. But for the United States, one dimension of the competition overshadows all others: the race with China. Beijing has set a goal of fielding one million tactical drones. Washington is restructuring its procurement system after the Replicator initiative fell short of its targets. Both nations are making bets, on AI quality versus manufacturing scale, on autonomy versus attritable mass, on the technological frontier versus the production floor. The race to master autonomous warfare will determine the balance of military power for the next generation. And the United States is not clearly winning.

Chapter 11: The Race

PART III: THE STAKES

In January 2026, inside a conference room at the Defense Innovation Unit's Mountain View headquarters, the building that Pentagon insiders call "the Zoom," because its glass-walled meeting spaces face directly onto Moffett Federal Airfield, where experimental aircraft have been taking off since the 1930s, a program manager named nothing that would appear in any press release pulled up three screens for a briefing that captured the central dilemma of American military power.

On the first screen: a slide deck for something called the Orchestrator Prize Challenge. The Defense Autonomous Working Group (DAWG) (the successor to the Pentagon's failed Replicator initiative) had partnered with DIU to launch a $100 million, six-month competition. The objective was deceptively simple and staggeringly ambitious: build AI software that can command drone swarms across air, ground, and maritime domains through voice and text commands. Speak to the system in plain English. Tell it what you need. The AI interprets the command, fuses data from every platform in the fleet in real time, and executes, all within human-defined constraints. Among the participants: SpaceX and xAI, Elon Musk's companies, alongside the usual defense-tech competitors. The outcome was expected by summer 2026.

On the second screen: a list of those competitors, startups, defense primes, and commercial technology companies, each betting that their software could crack the command-and-control problem that had eluded military planners for a decade. The participants ranged from established defense-tech firms already building swarm-capable platforms to commercial AI companies that had never touched a military contract. The competition was designed to be open, fast, and ruthless. Six months. One winner. No acquisition bureaucracy in between.

On the third screen: satellite imagery of Chinese drone production facilities. Assembly lines running at a tempo the United States could not match. Warehouses stacked with tactical unmanned aircraft systems (UAS) rolling off production floors that operated on a commercial manufacturing cadence, not a defense acquisition timeline. The imagery was not classified. It did not need to be. The scale was visible from space. One facility alone appeared to be producing more drones per month than the entire Replicator initiative had fielded in two years.

The juxtaposition was not lost on anyone in the room. The Orchestrator Challenge represented the sharpest edge of American defense innovation—a bet that the right software, built by the right company, could multiply the combat power of whatever drones the US managed to field. The Chinese production imagery represented the blunt reality of the competition: software multiplied by zero is still zero, and you cannot orchestrate drones you do not have.

The Orchestrator Challenge was the Pentagon's latest attempt to solve a problem that had haunted American defense planning since the Replicator initiative fell short of every target it set: how to field autonomous systems at scale before China does.

And the three screens in that conference room told the story in a single glance, American ingenuity on one side, Chinese production volume on the other, and a $100 million prize in the middle trying to bridge a gap that might not be bridgeable with software alone.

The Replicator Failure

The story of how the United States arrived at this moment begins with an announcement that was supposed to change everything.

On August 28, 2023, Deputy Secretary of Defense Kathleen Hicks stood at the National Defense Industrial Association's Emerging Technologies conference and unveiled the Replicator initiative. The ambition was deliberate and public: field "multiple thousands" of attritable autonomous systems across multiple warfighting domains by August 2025. Two years. Thousands of expendable drones, unmanned surface vessels, and ground robots, delivered at a speed that the Pentagon's glacial procurement system had never achieved. Hicks framed it as a direct response to China's military buildup, a bet that the United States could use autonomy and mass to offset Beijing's advantage in sheer numbers.

The initiative carried institutional credibility. Hicks was the second-highest-ranking civilian in the Department of Defense. The timeline was specific. The quantity was stated. The strategic logic was sound: if the future of warfare favored cheap, expendable, AI-enabled platforms over expensive crewed systems, then the United States needed to build them fast and field them faster.

Two years later, the results were in. Replicator had fielded hundreds of systems, not thousands.

Hundreds. Not the "multiple thousands" that Hicks had promised. Not even close. The initiative that was supposed to demonstrate that the United States could match China's production tempo had instead demonstrated the opposite: that even with the direct backing of the deputy secretary of defense, the Pentagon's acquisition system could not deliver expendable systems at scale within a two-year window.

The gap between promise and delivery was not a technology problem. The drones existed. Companies like Skydio, Shield AI, and Anduril had platforms ready to ship. The commercial drone market was producing millions of units per year for civilian customers. The failure was institutional, a procurement system designed to buy small numbers of exquisite weapons over multi-year timelines, asked to buy large numbers of expendable platforms in months. The contracting offices could not move fast enough. The testing and evaluation pipelines could not clear systems fast enough. The budget authorities could not reprogram funds fast enough. The authorization-to-operate process, designed to ensure cybersecurity compliance for million-dollar weapons platforms, was applied with equal rigor to $500 drones. The ambition was there. The bureaucracy was not.

On August 27, 2025 (almost exactly two years after Hicks announced Replicator) Secretary of Defense Pete Hegseth issued a memo that effectively killed the initiative in its original form. The program was restructured, renamed the Defense Autonomous Working Group, and its second phase consolidated into Joint Interagency Task Force 401 (JIATF-401). Hegseth's broader "Drone Dominance" campaign shifted emphasis toward equipping small military units with smaller systems, particularly first-person-view drones, the same class of weapon that had proven decisive in Ukraine. Meanwhile, the Pentagon's newly created DOGE unit was leading a parallel effort to acquire at least 30,000 cheap drones, a number that sounded large until measured against China's million-drone target. The first acquisition under the DAWG framework came on January 11, 2026: Fortem Technologies' DroneHunter F700, an AI-driven interceptor drone for counter-UAS missions. Not thousands of offensive systems. A single counter-drone platform.

The Replicator failure was not unique. It was symptomatic. The pattern repeats across the defense establishment: ambitious goals set by senior leaders, colliding with a procurement infrastructure that was designed for the Cold War and has never been fundamentally reformed. The United States does not lack the technology to compete in the autonomous systems era. It lacks the institutional capacity to buy that technology at the speed the threat demands. And while the Pentagon was restructuring its initiative, renaming its working group, and selecting its first counter-drone interceptor, China was building a million drones.

The Million-Drone Program

Beijing's approach to the autonomous weapons competition starts from a fundamentally different premise than Washington's.

The United States treats AI military adoption as a technology problem: invent the best algorithms, build the most capable platforms, and deploy them through a reformed acquisition system. China treats it as a diffusion problem: take existing technology (much of it commercially developed, some of it borrowed, all of it good enough) and push it into the military at scale through a system designed to eliminate the friction between civilian innovation and military demand.

The system is called Military-Civil Fusion (MCF), and it is not a slogan. It is an ecosystem directed from the top by President Xi Jinping: state-guided universities and research laboratories, a web of dual-use commercial vendors, and procurement pipelines that can route a capability from a Shenzhen startup to a People's Liberation Army (PLA) unit with a speed that Pentagon acquisitions officers can only envy. A Georgetown Center for Security and Emerging Technology (CSET) study analyzed 2,857 PLA AI-related contract award notices from January 2023 through December 2024 and found a defense industrial base anchored by state-owned enterprises but increasingly complemented by a long tail of nontraditional commercial vendors feeding directly into military demand.

The numbers tell the story. Beijing has launched a program to field one million tactical UAS by 2026, a target that, even if only partially achieved, would dwarf any Western drone production effort by an order of magnitude. In 2025, the PLA revealed a 200-drone AI swarm controlled by a single soldier, using what Chinese military media described as an "intelligent algorithm" that allowed the drones to cooperate autonomously even after losing communication with the operator.

The hardware is equally ambitious. On December 11, 2025, the Aviation Industry Corporation of China (AVIC) completed the maiden flight of the Jiu Tian, a drone mothership that the Western press immediately labeled the most ambitious swarm delivery platform in existence. The numbers: 16.35 meters long, 25-meter wingspan, 16-metric-ton maximum takeoff weight,

6,000-kilogram payload capacity, 7,000-kilometer ferry range, 12-hour endurance. The Jiu Tian carries approximately 100 small drones in dual internal bays, smart quadcopters and winged attack drones designed to be released en masse. Concept videos released by AVIC showed Jiu Tians launching swarms of hundreds of autonomous weapons to overwhelm US Navy carrier groups.

The China Electronics Technology Group Corporation (CETC) has tested fixed-wing drone swarms that go further still. CETC's Swarm I debuted at the 2021 Zhuhai air show; Swarm II, upgraded to 100 kilometers per hour and more than one hour of endurance, followed in 2024. Rather than relying on a central controller (the single point of failure that makes most current drone operations vulnerable) CETC swarms use onboard algorithms for real-time pathfinding, automated target recognition, and multi-agent reinforcement learning. Each drone processes its own sensor data, communicates with its neighbors, and makes decisions locally. Cut the link to the ground station, and the swarm keeps operating.

The China Aerospace Science and Technology Corporation (CASC) rounds out the triad, producing the CH-series drones, the CH-4, the CH-5, for both domestic use and an aggressive export market that has put Chinese military drones in the arsenals of nations across Africa, the Middle East, and Central Asia. China is not just building autonomous weapons for itself. It is building them for the world, creating dependency relationships that mirror what DJI built in the commercial market, and proving far harder to counter.

This is the diffusion model in action. China did not invent multi-agent reinforcement learning. Chinese AI researchers published much of their foundational work in the same academic journals as their American counterparts. The underlying science is largely shared. But China built the institutional machinery to move that research from a laboratory paper to a fielded military system faster than the United States could move a contract through its acquisition pipeline. The PLA's modernization strategy describes three overlapping phases: mechanization, informatization, and what Chinese military writing calls "intelligentization", the integration of artificial intelligence into every aspect of military operations. The autonomous weapons programs described above are the leading edge of that third phase. And the pace is accelerating.

Quality Versus Quantity

The American response to China's production advantage is a bet on quality.

The Collaborative Combat Aircraft (CCA) program represents the most expensive and most ambitious expression of that bet. The US Air Force plans to spend $8.9 billion on CCA development between fiscal years 2025 and 2029, building AI-powered autonomous drones that will fly as "loyal wingmen" alongside crewed fighter jets. In April 2024, the Air Force selected General Atomics and Anduril Industries as the two finalists for Increment 1. By March 2025, the competing prototypes had designations: the General Atomics YFQ-42A and the Anduril YFQ-44A, nicknamed "Fury." The YFQ-42A made its maiden flight in August 2025. The YFQ-44A followed on October 31, 2025.

The target: 1,000 CCAs at $25 million to $30 million each, a fraction of the $80 million price tag of an F-35 Lightning II, but still a $25 billion to $30 billion fleet. These are not attritable systems. They are sophisticated, semi-autonomous platforms designed to serve as sensor extensions, electronic warfare nodes, weapons carriers, and decoys for the crewed fighters they accompany. Air Force Secretary Frank Kendall envisioned up to five autonomous CCAs operating alongside a single next-generation fighter, transforming the human pilot from a dogfighter into a mission commander directing a team of robotic aircraft.

That next-generation fighter now has a name. On March 21, 2025, President Trump announced the F-47, Boeing's sixth-generation air superiority fighter, awarded under a contract worth more than $20 billion. Air Force leaders plan to buy 185 or more F-47s, each with a combat radius exceeding 1,000 nautical miles and a top speed above Mach 2. The first flight is expected in 2028, with the aircraft becoming operational by 2029. The F-47 was designed from inception to operate with CCA wingmen. In the near term, F-22 Raptors are being upgraded with tablet-based control systems to serve as the first airborne CCA controllers, starting in fiscal year 2026.

The investment is real. The hardware is impressive. The concept of operations, a human mission commander directing a team of AI-powered wingmen in contested airspace, represents a genuine doctrinal innovation. And the Marine Corps is running a parallel track: in January 2026, Northrop Grumman and Kratos won a $231.5 million contract to operationalize the XQ-58 Valkyrie as a loyal wingman for Marine aviation, adding another service's demand to the CCA pipeline.

Here is the problem: the American quality bet and the Chinese quantity bet are not competing on the same terms.

The United States is building a force designed around networked manned-unmanned teams, a small number of exquisite crewed fighters, each commanding a handful of sophisticated autonomous wingmen, operating in contested airspace against a peer adversary. The concept is sound. The technology is real. The timeline stretches to 2029 and beyond.

China is building a force designed around mass. A million tactical drones. Motherships carrying a hundred at a time. Swarms that operate without central control. The individual systems are less sophisticated than a CCA. They cost less. They are more expendable. And there are far, far more of them. The Pentagon's own wargames have tested this dynamic: Air Force planners designing the CCA Increment 2 originally envisioned a high-end, stealthy platform, but simulations of a Pacific conflict showed that large numbers of low-end aircraft would be more effective than small numbers of high-end versions. The wargames forced a rethink, with the Air Force futures directorate leaning toward a cheaper, mass-producible design for Increment 2.

The lesson of Ukraine applies here with uncomfortable directness. In the Donbas, both sides discovered that quantity has a quality all its own, that 10,000 $200 FPV drones create a problem that expensive counter-drone systems struggle to solve, not because any individual drone is capable but because the swarm saturates every defense. Ukraine produced more than 3 million drones in 2024 and projected 7 million for 2026, and Ukraine is a country fighting for its survival with a fraction of China's industrial capacity. If the Taiwan Strait becomes the next proving ground for autonomous warfare, the question is whether 1,000 CCAs at $25 million each can defeat a million drones at a fraction of the cost.

The honest answer is that nobody knows. The technology has never been tested at that scale, in that environment, against that adversary. But the trajectory favors the side that can produce more, faster, cheaper, and right now, that side is not the United States.

The numbers sharpen the point. At current production rates, the United States will have approximately 1,000 CCAs operational in the early 2030s if the program stays on schedule, a significant assumption given the history of major defense programs. China's million-drone target, even if only half-achieved, would yield a 500-to-1 numerical advantage in tactical unmanned systems by that same date. The US platforms will be smarter, more survivable, and more capable per unit. The Chinese platforms will be cheaper, more expendable, and far more numerous. History offers no clean precedent for resolving this kind of asymmetry. The closest analog may be the Eastern Front in World War II, where Soviet T-34 tanks (simpler, cheaper, and produced in vastly greater numbers) overwhelmed the technically superior German Panthers and Tigers. Quality won individual engagements. Quantity won the war.

The Decoupling

The US-China drone competition is not only a contest of production and technology. It is also a contest of supply chains, and in 2025, those supply chains began to come apart.

DJI, the Shenzhen-based company that controls more than 70% of the global consumer drone market, had been under legislative scrutiny for years. The Countering CCP Drones Act, proposed to ban DJI outright from US markets, did not pass as standalone legislation. Instead, Congress inserted a "review trigger" into the 2025 National Defense Authorization Act (NDAA): if a federal agency did not complete a formal security audit of DJI by December 23, 2025, the company would be automatically added to the Federal Communications Commission's (FCC) Covered List.

No audit was completed. On December 22, 2025, the FCC acted, and went further than anyone expected. Rather than targeting DJI specifically, the commission added all foreign-made unmanned aircraft systems and critical UAS components to its Covered List, based on a national security determination that these products "pose unacceptable risks to the national security of the United States." The placement prevents any new foreign-made drone model from receiving FCC equipment authorization. Without FCC authorization, a drone cannot be legally imported, marketed, or sold in the United States. Existing models that had already received approval could continue to operate. But the pipeline of new Chinese drones into the American market was shut.

The Autel blacklisting had come first. On January 6, 2025, the Department of Defense added Autel Robotics (headquartered in Shenzhen and previously considered a viable alternative to DJI for US government customers) to its blacklist as a "Chinese Military Company." The two largest Chinese drone manufacturers were now both locked out of the American defense and government market.

China retaliated. On April 9, 2025, China's Ministry of Commerce placed Shield AI, Skydio, and BRINC Drones on its Unreliable Entities List, barring them from sourcing parts, operating, investing, or visiting China. Shield AI's president, Brandon Tseng, responded publicly that the move would not affect the company because it was "intentionally not reliant on China." But Skydio's experience told a more complicated story. In October 2024, Chinese sanctions had blocked Skydio's access to its sole lithium-ion battery supplier, Dongguan Poweramp, a TDK subsidiary, forcing the company to ration batteries to one per drone until it could establish a domestic supply chain. China controls more than 70% of the global battery market. "Intentionally not reliant on China" was a strategic aspiration, not a current reality, for much of the American drone industry.

The decoupling is now mutual, accelerating, and irreversible. The United States is building a drone industrial base that excludes Chinese components. China is building a drone industrial base that excludes American customers. The global drone market ($47 billion in 2025, projected to reach $98 billion by 2033) is splitting into two ecosystems, each with its own supply chains, its own standards, and its own strategic logic. The parallel is not to the semiconductor export controls of 2022, which attempted to restrict Chinese access to specific chokepoints in the chip supply chain. The drone decoupling is broader: it is an attempt to build an entirely separate industrial base from raw materials to finished platforms, in a domain where China has spent two decades establishing dominance at every level of the stack.

For the drone operators, first responders, and commercial pilots who depend on DJI products, the impact is immediate. Roughly 43% of commercial drone pilots surveyed anticipated "extremely negative" or "business-ending" consequences from the ban. More than 80% predicted they would be out of business within two years without access to new DJI models. The price gap compounds the pain: a Skydio X10 costs $15,000 to $20,000, compared to approximately $5,000 for a DJI Mavic 3. The three-to-four-times price premium is the cost of decoupling, and someone has to pay it.

The Taiwan Variable

Every calculation here (every production target, every CCA prototype, every supply chain decision) is shaped by a scenario that neither Washington nor Beijing discusses publicly with candor: a conflict over Taiwan.

The intelligence community's assessment, as stated in the Defense Intelligence Agency's 2026 threat report, is measured: Chinese leaders do not currently plan to execute an invasion of Taiwan in 2027, nor do they have a fixed timeline for achieving unification. Beijing publicly insists that unification with Taiwan is required to achieve its goal of "national rejuvenation" by 2049. Chinese officials recognize that an amphibious invasion would be extremely challenging and carry a high risk of failure, especially in the event of US intervention.

But the preparations are real. China is making steady progress on capabilities it would use in any attempt to seize the island, and the autonomous systems described above would feature prominently in that scenario. The Jiu Tian mothership was not built for parades. CETC's decentralized swarms were not designed for air shows. The million-drone program is not an industrial exercise. These are systems designed to saturate a defensive perimeter, overwhelm air defenses, and create the conditions for an amphibious crossing of 100 miles of open water against the opposition of the most powerful navy on Earth.

The autonomous weapons dimension of a Taiwan scenario operates on three levels.

First, anti-access and area denial. China's strategy for the western Pacific depends on keeping US carrier groups far enough from Taiwan that their aircraft cannot provide effective support. Drone swarms (launched from motherships, from surface vessels, from the Chinese mainland itself) offer a mechanism for generating the kind of distributed, sustained threat that traditional anti-ship missiles cannot. A carrier group's air defenses are designed to intercept a finite number of high-value threats, cruise missiles, manned aircraft. They are not designed to intercept hundreds of autonomous drones arriving simultaneously from multiple vectors at varying altitudes. The math favors the attacker. A Jiu Tian carrying 100 drones costs a fraction of a single anti-ship cruise missile. A dozen Jiu Tians carrying 1,200 drones would present a targeting problem that no existing naval air defense system has been tested against.

Second, the counter-swarm problem. Taiwan received its first tranche of American-made Altius loitering munitions in August 2025, expendable drones that can be rapidly deployed across multiple platforms to serve as a defensive screen. The United States is investing $7.5 billion in counter-UAS systems in fiscal year 2026. But the fundamental asymmetry persists: building drones is cheaper and faster than building the systems to defeat them. Counter-swarm defense requires diverse, integrated sensors to track small, low-flying targets; layered effectors to progressively thin incoming swarms; and AI battle management systems that match threats to effectors within seconds. No military has demonstrated this capability at scale against a determined peer adversary.

Third, communications and resilience. A Taiwan Strait conflict would be saturated with electronic jamming, GPS denial, and cyberattacks, the same electronic warfare environment that has shaped every dimension of autonomous weapons development. Taiwan's defense depends on spotting Chinese amphibious forces early and striking them fast. In a conflict, both Taiwan and the United States would rely on commercial radar satellites and communications constellations to sustain dispersed forces and coordinate strikes across the Pacific. Beijing understands this. It is building jammers, directed-energy weapons, and cyber tools aimed not only at US military satellites but at the commercial networks Washington depends on. Autonomous systems that can navigate without GPS, communicate through mesh networks, and execute mission objectives without continuous human guidance (the exact capabilities that Hivemind, Lattice, and the Skydio Autonomy Engine were designed to provide) become not optional but essential.

The integration of AI into this scenario introduces a danger that extends beyond the tactical. Autonomous systems operating at machine speed (selecting targets, coordinating strikes, and adapting to countermeasures faster than any human commander can process) compress the decision cycle to a tempo where meaningful human oversight becomes physically impossible. In a Taiwan Strait engagement involving thousands of autonomous platforms on both sides, the risk is not that a commander makes a bad decision. The risk is that the systems on both sides interact in ways no commander anticipated, producing escalatory spirals at a speed that outruns human control. The brittleness of algorithmic systems, their potential for unpredictable behavior at the edges of their training data, and the opacity of their decision-making processes all compound a risk that defense planners on both sides acknowledge privately but rarely discuss publicly: that autonomous weapons could escalate a Taiwan crisis faster than either government can de-escalate it.

The Taiwan scenario is not a prediction. It is the strategic context that explains why both nations are investing at the scale and speed described above. It is the reason China is building a million drones and the United States is spending $8.9 billion on autonomous wingmen. It is the reason the DJI ban and the supply chain decoupling are national security imperatives, not trade disputes. And it is the reason the Orchestrator Prize Challenge (a $100 million competition to build AI that can command swarms through voice commands) is not a technology demonstration but an urgent attempt to solve a problem that may determine the outcome of the most consequential military confrontation of the 21st century.

The Race Neither Side Can Afford to Lose

The US-China competition in autonomous military systems is the most consequential arms race since the nuclear era. But unlike the nuclear arms race, where both sides eventually reached a rough parity enforced by mutual assured destruction, the autonomous weapons race has no natural equilibrium. There is no equivalent of a nuclear warhead count that both sides can verify. There is no treaty framework, the UN General Assembly vote in December 2024, in which 166 nations supported autonomous weapons regulation, produced a mandate for negotiation, not a binding agreement. There is no red phone.

The asymmetry is structural. The United States has better AI. Its commercial technology sector (the companies building the foundation models, the edge computing chips, the autonomy software) is years ahead of China in the quality of its artificial intelligence. Hivemind can fly a drone through a building with no

GPS, no communications, and no human input. Lattice can fuse sensor data across domains in real time and present a commander with a coherent operational picture. The Skydio Autonomy Engine can navigate a GPS-denied environment using six cameras and a neural network that builds a three-dimensional map of the world in real time. These capabilities are real, demonstrated, and deployed. China has nothing equivalent in the public record.

But China has more drones. It has more production capacity. It has an institutional system, Military-Civil Fusion, that moves technology from the commercial sector to the military faster than any Western democracy can. And in the emerging era of autonomous swarm warfare, where the decisive variable may not be the intelligence of any individual platform but the number of platforms a military can put in the air simultaneously, production volume may matter more than technological sophistication.

The Pentagon's AI Acceleration Plan, released in early 2026, attempts to address this asymmetry directly. The strategy treats AI adoption as an operational race in which the decisive variable is diffusion, how quickly a promising capability moves from experiment to trusted, fielded use across the services and commands that actually fight. Rather than trying to replicate China's centralized, state-directed model, the plan uses "Pace-Setting Projects" combined with forced enterprise-wide replication to push AI capabilities through the bureaucracy. Every military department, combatant command, and defense agency is required to identify within 30 days at least three projects to fast-follow the initial demonstrations. A monthly "barrier removal board" is empowered to waive nonstatutory requirements and escalate the rest. The signal is explicit: compliance can no longer function as a veto over fielding timelines.

The strategy is serious. Whether it is sufficient is another question. The Pentagon has announced AI strategies before. The Joint Artificial Intelligence Center, established in 2018, was supposed to centralize and accelerate military AI adoption; it was reorganized into the Chief Digital and Artificial Intelligence Office in 2022. Project Maven, the drone-footage analysis program that triggered the Google employee revolt in 2018, was supposed to demonstrate that commercial AI could be rapidly integrated into military operations; eight years later, the military is still trying to solve the same integration problems at scale. The pattern is familiar: announce a bold initiative, staff it with capable people, watch it collide with the institutional antibodies of a $886 billion bureaucracy, and then announce the next bold initiative. DAWG is the latest entry in this cycle. The AI Acceleration Plan may break the cycle. Or it may become another memo in a filing cabinet.

The race cannot be won by the Pentagon alone. Traditional defense primes, Lockheed Martin, Raytheon, Northrop Grumman, were built for a world of exquisite, expensive platforms manufactured on decade-long timelines. Their revenue models depend on long-cycle programs of record, cost-plus contracts, and the kind of engineering complexity that generates decades of maintenance and upgrade revenue. The drone revolution demands something their business models were never designed to deliver: speed, software, attritable mass, and the willingness to build weapons designed to be lost. A $200 drone that is destroyed on its first mission does not generate a maintenance contract.

But the new defense-tech companies (the ones that emerged from Silicon Valley garages and Navy SEAL deployments and MIT robotics laboratories) were built for exactly this world. They move at commercial speed. They iterate in weeks, not years. They build software-defined platforms that can be updated overnight. And they are already producing the autonomous systems that the United States needs to compete. The question is whether the Pentagon can learn to buy from them at the speed and scale the threat demands. That question, and those companies, are the subject of Part IV.

Chapter 12: The American Drone

PART IV: THE PATH FORWARD

On a Friday afternoon in March 2026, the 550th quality checkpoint cleared and the X10D was done. A technician at Skydio's Hayward, California manufacturing facility lifted the drone from the line, turned it once in her hands, and set it on a pallet alongside 47 others that had cleared final inspection that day. Nine minutes earlier, the drone had been components, a folded airframe, six navigation cameras, an NVIDIA Jetson Orin GPU, a Teledyne FLIR Boson+ thermal sensor, a multiband radio, and the lithium-ion battery that had caused the company so much trouble the year before when China cut off supply. Now it was an autonomous robot that could deploy from a backpack in 40 seconds, navigate a GPS-denied battlefield without a single satellite signal, and build a three-dimensional map of its surroundings at more than one million points per second. The facility was producing over a thousand of them every month. It was not enough.

Three days before this particular drone cleared its 550th checkpoint, the US Army had placed a $52 million order for over 2,500 X10Ds, the largest single-vendor tactical small unmanned aircraft system order in Army history. The order moved through Atlantic Diving Supply, a procurement intermediary, and the contract traveled from bid to award in under 72 hours. Seventy-two hours. The same institution that routinely takes 18 to 36 months to move from a request for proposals to a first payment had purchased 2,500 autonomous drones in the time it takes to process a weekend's worth of email. The order had arrived three and a half weeks into the active US-Iran military conflict, where drones had become the decisive weapon class. The Army did not have time to wait.

The production floor itself occupied more than 70,000 square feet of a facility on Clearview Way in Hayward, across the bay from San Francisco. It did not look like a Lockheed Martin assembly line. There were no robotic arms welding fuselage sections, no clean rooms with gowned technicians handling stealth coatings. It looked more like a high-end electronics factory, organized workstations where technicians in company t-shirts assembled compact airframes, mounted cameras, loaded software, ran diagnostics. Over 200 workers cycled through the facility. The scale was modest by defense-industrial standards. It was the largest drone production operation in the United States.

Adam Bry walked the production floor that afternoon with the look of a man who had been proven right about something he wished he had been wrong about. The CEO and co-founder of Skydio had spent a decade arguing that America needed a domestic drone industrial base, that the country's dependence on Chinese-manufactured drones was, in his words, "a fragile and dangerous dependency" that an adversary could weaponize. In October 2024, China had proved him correct by imposing sanctions that blocked Skydio's access to its sole battery supplier, Dongguan Poweramp. By March 2025, China had placed Skydio on its export control list. The company that marketed itself as the American alternative to Chinese drones had nearly been killed by its own supply chain.

Now, a year later, the supply chain was domestic, the batteries were American-sourced, the Army order was the largest in history, and 29 allied nations depended on the drone clearing its final checkpoint on this production floor. On the wall near the entrance, a timeline charted the company's trajectory: Seed round, $3 million, 2015. Series A, $25 million, Andreessen Horowitz, 2016. Consumer market exit, August 2023. Series E Extension, $170 million, November 2024. Total raised: approximately $740 million. Revenue: $180 million in 2024, up 80% year over year. Bookings pipeline: $1.2 billion, with more than half from defense contracts. Valuation trending toward $3 billion on secondary markets.

Down the hall from the production line, a smaller team was finalizing a different kind of bet. The Skydio F10, the company's first fixed-wing platform, announced at the Ascend conference in September 2025, was entering early-access production. Ninety-plus minutes of flight time. Eighty-plus miles per hour. Dozens of miles of range. Robotic Takeoff and Landing that eliminated the need for a runway or a launch rail. If the X10D was a hunter that could see everything within a few kilometers, the F10 was a hunter that could see everything within a county. Bry had been talking publicly about the "1-to-many" vision for years (one operator managing five, 10, 50 drones simultaneously) and the F10 was the platform that would make that vision operational for the first time at range. Combined with the Skydio Dock, the automated base station that could launch, recover, and recharge drones autonomously, the F10 would give a single soldier persistent aerial surveillance over a battlefield-sized area without dedicating a single additional body to the task. The company's Drone as First Responder software had already surpassed 10 million calls for service in the law enforcement domain by February 2026. The military version of that capability, always-on, always-watching, always-autonomous, was what the F10 was built to deliver.

Skydio was not a startup anymore. It was the company the United States was betting on to replace Chinese drones across the military, across allied nations, and across domestic law enforcement. The question (the question that defined not just Skydio but the entire American drone industrial base) was whether it could scale fast enough.

The Three Pillars

The American drone industrial base of 2026 does not look like anyone predicted a decade ago. It is not dominated by the legacy defense primes, the Lockheed Martins, the Northrop Grummans, the General Atomics that built the Predator and Reaper fleets. Those companies still hold enormous contracts and institutional relationships. But the companies building the autonomous systems that the Pentagon actually needs for the drone age are younger, faster, and built on fundamentally different engineering philosophies.

Three companies define the emerging base. Each represents a different function on the Hunter-Killer spectrum, and together they form the skeleton of what American autonomous warfare will look like for the next generation.

Skydio is the hunter. The company described in Chapters 2 and 5, built on MIT robotics research, powered by the Autonomy Engine that runs nine deep neural networks simultaneously, deployed by every US military branch and 29 allied nations. Skydio makes the autonomous sensing platform: the drone that finds, maps, and identifies targets in environments where GPS is denied and electronic warfare has neutralized every other system. The X10D does not carry weapons. It carries eyes, eyes that see in thermal and visual spectra, that build three-dimensional maps in real time, that operate when satellites are jammed and radios are dead. The hunter half of the equation. Revenue over $180 million in 2024. Bookings pipeline of $1.2 billion. Approximately 900 employees. Valued at roughly $2.5 billion, trending toward $3 billion. The largest drone manufacturer in the United States.

Anduril is the mesh and the killer. The company described in Chapter 6, Palmer Luckey's defense startup that grew from zero to a $60 billion valuation in nine years by building Lattice, the AI software platform that fuses sensor data from every domain into a single operational picture. Anduril does not just connect systems. It builds the weapons those systems command: the Altius loitering munition, the Fury autonomous combat aircraft, the Anvil counter-drone interceptor, the Ghost X reconnaissance drone that feeds data directly into the mesh. The $20 billion Army counter-drone contract awarded in March 2026 was not a hardware purchase. It was a bet that the software layer connecting drones, sensors, and weapons into a coordinated network would determine who controls the battlefield. Anduril's revenue reached $2 billion in 2025. The company employed 7,000 people and was building Arsenal-1, a $1 billion hyperscale manufacturing facility in Ohio designed to produce autonomous systems at a pace that legacy factories cannot match.

Shield AI is the brain. The company described in Chapter 5, Brandon Tseng's startup born from a Navy SEAL's experience in Afghanistan, where the absence of autonomous reconnaissance cost his teammates' lives. Shield AI's Hivemind software is an autonomous pilot: artificial intelligence that flies aircraft without human input in GPS-denied, communications-denied environments. Where Skydio's autonomy navigates a small drone through a parking garage, Hivemind's ambition is to fly fighter jets. The company demonstrated Hivemind on US Navy combat aircraft in manned-unmanned teaming exercises. It operates the V-BAT Teams program, autonomous vertical-takeoff-and-landing drones that conduct ISR and strike missions in contested environments. Shield AI's valuation reached $5.6 billion in 2025 after raising $240 million in a Series F round. The company employed more than 1,000 people and maintained a Kyiv office for direct engagement with Ukrainian combat operations. Revenue hit $267 million in 2024, up 64% year over year.

Hunter, mesh, brain. Sensing, connecting, deciding. Three companies, three functions, three pieces of an architecture that the United States needs to field as an integrated system. The technology exists. The capital exists. The question is whether the institution that must buy it, the Department of Defense, can move fast enough to assemble the pieces before the next crisis makes the current one look like a rehearsal.

The Money

The capital that built these three companies did not come from the Pentagon. It came from Sand Hill Road.

Defense-tech startups raised a record $49.9 billion in 2025: nearly double the $27.2 billion raised the previous year, according to PitchBook data. Equity funding alone more than doubled, surging from $7.3 billion in 2024 to $17.9 billion in 2025, per CB Insights. The growth outpaced overall venture capital, which rose 47% in the same period on the back of AI investment. Defense tech was not riding the wave. It was the wave.

The numbers tell a story of radical realignment. A decade ago, mainstream venture capital firms treated defense investing as somewhere between distasteful and radioactive. The Google employee revolt over Project Maven in 2018,when thousands of Google engineers signed a petition demanding the company withdraw from a Pentagon contract to apply AI to drone surveillance footage, had established the cultural norm: building weapons was incompatible with the values of the technology industry. The number of VC firms willing to write checks to defense startups could be counted on two hands.

By 2025, the number of firms actively investing in defense tech had increased 41%, according to CB Insights. Mainstream venture capital dropped its ethical objections and reframed defense investing as "supporting democratic values." The firms leading the charge were not generalists dabbling in a new sector. Andreessen Horowitz built an entire practice, American Dynamism, around the thesis that investing in national security companies was both patriotic and profitable. Founders Fund, Peter Thiel's firm, wrote a $1 billion check for Anduril's Series G. Lux Capital and Shield Capital hired former defense officials and built Washington policy teams. Y Combinator, the startup accelerator that had launched Airbnb and Stripe, began funding missile companies.

The average defense-tech Series B in 2025 was $87 million, according to PitchBook, nearly double the $45 million average for consumer-tech Series Bs. The premium reflected the reality of building defense companies: hardware manufacturing costs more than software, security clearances take longer than code sprints, and passing ITAR compliance reviews requires lawyers that consumer startups do not need to hire. But the premium also reflected something more fundamental. Investors had discovered that defense contracts, once won, create the kind of durable competitive moats that had been absent from venture portfolios since the mobile platform wars ended. A DoD contract takes 18 to 36 months to win, but once awarded, the revenue is predictable for years. A competitor cannot undercut you with a price war or clone your product in six weeks. The compliance requirements: ITAR, CMMC 2.0, FedRAMP, NIST 800-171, function as barriers to entry that keep competitors out. Every dollar spent on regulatory qualification is a dollar a future competitor must also spend before they can even bid on contracts.

The exits proved the thesis was not speculative. VC exits from defense-tech investments jumped to $54.4 billion in 2025, up from $18.2 billion the previous year, driven largely by prime contractors acquiring proven startups to integrate into their legacy systems. The money was flowing in, generating returns, and attracting more money behind it.

The phenomenon was not uniquely American. In Europe, Helsing, a Munich-based battlefield AI company founded in 2021, reached a valuation of 12 billion euros, roughly $13.9 billion, after a 600 million euro Series D round in June 2025. European defense-tech deal count grew 67% in 2025, outpacing the 30% growth in the United States. The drone age was building a global defense-tech ecosystem, and for the first time since the early Cold War, the private sector was outpacing government laboratories as the primary engine of military innovation. The implications were profound: the companies that would determine American military capability in the next decade were being funded by venture capitalists, not Congressional appropriations.

But here is the number that matters most for the argument of this chapter: manufacturing-focused defense investments rose to $4.7 billion across 39 deals in 2025. The VC community had figured out what the Pentagon was still learning. After the invention phase, after the proof-of-concept demonstrations, the DARPA contracts, the initial deployments, the battleground shifts to production. The companies that can convert prototypes into repeatable, high-volume manufacturing output will capture both capital and contract velocity. The companies that cannot will become footnotes.

Arsenal of Democracy, Version 2.0

The manufacturing challenge facing the American drone industrial base is not abstract. It is arithmetic.

Skydio's Hayward facility produces more than 1,000 drones per month. The facility employs over 200 manufacturing workers across its 70,000-plus square feet, and by March 2026, Skydio had produced more than 55,000 drones in total, a number that sounds significant until you place it against the scale of what the defense establishment is asking for. That is impressive for a company that was building consumer camera drones five years ago. It is not impressive compared to what the threat demands. The Pentagon's DOGE-led initiative is seeking to acquire at least 30,000 small drones. At Skydio's current production rate, filling that order alone, setting aside the Army's existing 2,500-unit order, NATO's framework agreement, and contracts with 29 allied nations, would take more than two years. Ukraine produced over 4 million drones in 2024 and is projected to build 7 million in 2026. China's drone production capacity is measured in millions of units annually across hundreds of manufacturers.

Anduril's response to the manufacturing gap is the most ambitious in the American defense-tech ecosystem. Arsenal-1, the company's hyperscale production facility under construction in Pickaway County, Ohio, represents a $1 billion bet that autonomous weapons can be manufactured at scale in the United States. The facility is designed to produce autonomous aerial and maritime systems (drones, loitering munitions, autonomous surface vessels) natively equipped with Lattice software, meaning every unit that rolls off the line is already integrated into the mesh. Arsenal-1 is not just a factory. It is a production philosophy: build the software first, then manufacture hardware that the software already knows how to command. The approach inverts the traditional defense-manufacturing model, where hardware is built first and software is bolted on later, often years later, at enormous cost, with results that rarely satisfy the original specification.

Shield AI's manufacturing ambitions are centered on the V-BAT platform and the scaling of Hivemind across multiple airframe types. The company's strategy is not to build thousands of identical drones but to build the AI pilot that can fly any aircraft, and then license or integrate that software across the broadest possible fleet. If Hivemind can autonomously pilot an F-16, it can autonomously pilot a CCA. If it can fly a V-BAT, it can fly any vertical-takeoff-and-landing platform the military procures. The brain scales differently than the body. Software replicates at near-zero marginal cost. Hardware does not.

The three approaches, Skydio's high-volume sUAS production, Anduril's hyperscale manufacturing with embedded software, Shield AI's software-first strategy across multiple platforms, represent different theories of how to build an American drone industrial base. The common thread is that none of them resemble how Lockheed Martin builds an F-35. The F-35 program took 24 years from contract award to full-rate production. It cost $1.7 trillion over its projected lifetime. It produced a fighter jet that the Government Accountability Office found had, as of 2024, hundreds of unresolved performance deficiencies. The legacy model builds expensive, exquisite platforms in small numbers over long timescales. The drone age demands cheap, attritable systems in massive volumes at wartime speed. The business models are incompatible.

The 72-Hour Standard

The $52 million Skydio order that moved from bid to award in 72 hours should not be remarkable. It should be normal.

The traditional Pentagon procurement timeline runs like this: a requirement is identified, a request for proposals is drafted and published, vendors respond over a period of weeks or months, proposals are evaluated, a contract is negotiated, legal review is completed, the award is made, and production begins. For a major program, this process takes years. The F-35 program launched in 2001 and achieved initial operational capability in 2015, 14 years from contract to fielding. The Army's Future Vertical Lift program, which aimed to replace the Black Hawk helicopter, ran for more than a decade before producing its first prototype. Even the Replicator initiative, launched by Deputy Secretary of Defense Kathleen Hicks in August 2023 with the explicit goal of fielding autonomous systems "at speed and scale", managed to deliver only hundreds of systems by its target date of August 2025, not the thousands it promised. The technology existed. The production capacity existed. The procurement system could not move fast enough to connect supply with demand.

The Skydio order bypassed that system. It moved through an Other Transaction Authority pathway, the procurement mechanism that allows the Pentagon to work outside the Federal Acquisition Regulation, the primary rulebook governing federal contracts. OTAs are not new. They have existed since the 1950s. But their use has accelerated dramatically as the Pentagon recognized that the FAR's 2,000-plus pages of rules, designed to ensure fairness and accountability in peacetime procurement, functioned as a chokepoint when speed was the operational requirement. OTA contract value grew 35% year over year in fiscal year 2024, according to the Government Accountability Office. The Skydio order demonstrated what OTA procurement looks like when it works: a requirement identified, a known vendor with a proven product, a contracting vehicle that skips the multi-month evaluation cycle, and drones on pallets within weeks.

The current administration has signaled that 72 hours should be the aspiration, not the anomaly. Secretary of Defense Pete Hegseth's January 2026 remarks framed the strategic competition as "a race in which the fastest innovator and iterator wins," rejecting what he called the Pentagon's "linear" acquisition model. The AI Strategy for the Department of War, released in early 2026, directed the department to become an "AI-first" warfighting force and identified bureaucratic friction, not technology gaps, as the primary obstacle. The strategy established a monthly Barrier Removal Board empowered to waive non-statutory procurement requirements and escalate impediments directly to the Deputy Secretary. It replaced the

Defense Innovation Steering Group, the Defense Innovation Working Group, and the CTO Council with a single CTO Action Group designed to "make decisions, clear bureaucratic blockers, and hold leaders accountable." The Under Secretary of War for Research and Engineering was designated as the department's single Chief Technology Officer with decision authority.

The reforms also target the data infrastructure that AI systems require. The AI Strategy directed the Chief Digital and Artificial Intelligence Officer to enforce the DoD's Data Decrees, requiring every military department and component to establish federated data catalogs and expose their system interfaces within 30 days. Intelligence data was to receive parallel treatment. The signal was unmistakable: data would no longer be hoarded by individual services and agencies as bureaucratic fiefdoms. It would be shared, standardized, and made available to AI systems across the department. Cameron Stanley, the newly appointed Chief Digital and AI Officer, called on the private sector to partner with the Pentagon on AI efforts, specifically encouraging the use of OTAs to reach nontraditional defense companies.

The reforms are structural, and they address real problems. But the test is not whether the Pentagon can write a memo that says "move faster." The test is whether the system can execute 72-hour procurement cycles repeatedly, across multiple programs, for companies that do not have Skydio's established relationship with the Army. The Skydio order moved fast because Skydio had spent four years building trust through the SRR program, had a Blue UAS-cleared product already in the Army's inventory, and had a contracting vehicle already in place through Atlantic Diving Supply. The next company, the defense-tech startup with a proven product but no existing Army relationship, no Blue UAS clearance, no established contracting pathway, will face every friction point the Skydio order avoided.

The Last Mile

The Pentagon's acquisition reform agenda contains one provision that, if executed, could matter more than all the others combined. Beginning in fiscal year 2028, each portfolio acquisition executive must establish an Innovation Insertion Increment, a dedicated funding pool to pay for rapid capability upgrades and modular insertions into existing programs. The provision addresses what defense-tech founders call the "last mile" problem: the gap between a successful demonstration and actual deployment.

The last mile kills companies. A defense-tech startup builds a product, demonstrates it at a military exercise, receives enthusiastic feedback from operators, wins a small initial contract, and then waits. Waits for the program office to secure follow-on funding. Waits for the test-and-evaluation bureaucracy to certify the system. Waits for an Authority to Operate. Waits for the contracting officer to work through the procurement regulations. One defense software company spent $4 million and 18 months getting its first DoD Authority to Operate. The delay was not caused by the technology failing. It was caused by the bureaucracy succeeding, at its own internal logic of process compliance, risk mitigation, and procedural completeness.

The Replicator initiative's shortfall is the definitive case study. When Hicks launched Replicator in August 2023, the stated goal was to field thousands of autonomous systems within two years. The technology was not the constraint. Skydio, Anduril, Shield AI, and dozens of smaller companies had products ready to deliver. The constraint was procurement. The initiative has since been renamed the Defense Autonomous Working Group, and its ambitions have been recalibrated. But the lesson stands: the United States does not have a technology problem. It has a buying problem. The drones exist. The software exists. The manufacturing capacity exists, or can be scaled with the right contracts. What does not exist, consistently, is the institutional ability to convert a requirement into a purchase order at the speed the threat demands.

The Innovation Insertion Increment is designed to fix that gap, to create a standing pool of money that program managers can use to buy new capabilities without waiting for a new budget cycle, a new authorization, or a new round of Congressional approval. Whether it works will depend on whether program executives treat it as a tool or a line item, whether they use it to actually buy systems at speed or whether it becomes another budget category that gets reprogrammed, deferred, or consumed by overhead.

The Defense Innovation Unit has been elevated to help bridge the gap from the other direction, not waiting for program offices to discover what the private sector has built, but actively identifying commercial capabilities and inserting them into military programs. The two-channel model the Pentagon envisions (DIU pulling commercial products in, the CTO Action Group pushing procurement barriers out) is architecturally sound. It acknowledges that the defense-tech ecosystem has outrun the Pentagon's ability to absorb its products, and that the solution requires institutional reform, not more technology demonstrations.

What Must Be True

The argument of this chapter reduces to three propositions.

The first is that the American drone industrial base already exists in embryonic form. Skydio, Anduril, and Shield AI are not the only companies, Saronic is building autonomous maritime vessels at a $4 billion valuation, Chaos Industries is building counter-drone radar at $4.5 billion, Rebellion Defense is building AI decision software, and dozens of smaller startups are entering the ecosystem every quarter. But the three pillars represent the core functions: autonomous sensing, networked command, and AI-driven decision-making. The $49.9 billion in venture capital that flowed into defense tech in 2025 is building a parallel defense industry, one that moves at Silicon Valley speed, builds software-defined systems, and measures success in months, not decades.

The second proposition is that the Pentagon's acquisition system is the binding constraint. Not technology. Not capital. Not talent. The 72-hour Skydio order proved that the system can move at wartime speed when the institutional will exists. The Replicator shortfall proved that it usually does not. The reforms announced in 2026,the Barrier Removal Board, the CTO Action Group, the Innovation Insertion Increment, the DIU elevation, the OTA expansion, are necessary. They are not sufficient. Sufficiency requires that the reforms survive contact with the bureaucratic immune system that has digested every previous reform effort. The Pentagon has been trying to reform acquisition since the Packard Commission in 1986. Forty years of reform efforts have produced incremental improvements and no fundamental change. The drone age does not have forty years.

The third proposition is that manufacturing scale is the next competitive battleground. The invention phase is over. The proof-of-concept phase is over. Skydio, Anduril, and Shield AI have demonstrated that American companies can build autonomous systems that match or exceed anything China or Russia has fielded. The question is whether they can build them in the volumes the threat requires. Ukraine produces millions of drones per year in a wartime economy. China produces millions in a peacetime one. The United States produces thousands. Skydio's 55,000 total drones produced since its founding represent roughly one day's output for the Chinese drone industry. The gap between thousands and millions is not a technology gap. It is an industrial-policy gap, a gap that Arsenal-1, the Hayward facility, and their successors must close before the next conflict makes the arithmetic unforgiving.

The economics point in one direction. An X10D costs approximately $17,300. An FPV drone built from commercial components costs $200. A Switchblade 600 loitering munition costs roughly $50,000. The attritable thesis (the concept that autonomous systems should be cheap enough to lose without strategic consequence) requires production volumes that drive unit costs down. Skydio's pricing reflects low-volume manufacturing of a sophisticated platform. At 10,000 units per month instead of 1,000, the cost curve bends. At 100,000 per month, it transforms. Whether American manufacturers can reach those volumes depends on contracts large enough to justify the capital expenditure, and contracts that large depend on an acquisition system willing to commit to volume purchases. The chicken-and-egg problem is real, and it has no purely market-based solution. It requires the Pentagon to act as the customer of last resort, to place the large, sustained orders that give manufacturers the confidence to invest in the production capacity the threat demands.

Underlying all three propositions is a talent dynamic that cuts both ways. The defense-tech ecosystem's greatest asset is the people it has attracted from both Silicon Valley and the Pentagon. Brose left the Senate Armed Services Committee for Anduril. Tseng left the SEALs for Shield AI. The Defense Innovation Unit, created to bridge the Pentagon and the technology industry, has seen its alumni populate dozens of defense-tech startups. Skydio's own chief legal officer, Brendan Groves, came from the Department of Justice, where he had managed the national security policy portfolio. The talent pipeline flows from government into the private sector because the private sector offers equity, speed, and the ability to build rather than brief. But the pipeline also hollows out the institutions that must buy what these companies build. Every procurement officer who leaves for a startup is one fewer person inside the Pentagon who understands how to move a contract at speed. Every DARPA program manager who departs for venture capital is one fewer person who can identify which startup's technology is ready for fielding and which is a PowerPoint presentation with a valuation. The ecosystem that is building America's autonomous future is simultaneously draining the institution that must field it.

Skydio, Anduril, Shield AI. Hunter, mesh, brain. The pieces exist. The capital exists. The technology works. The remaining variable is the institution that must assemble them into a force, and whether that institution can overcome four decades of its own inertia before the window closes.

Building the drones is the first challenge. Governing them is the second, and harder, one. The same technology that gives the United States a military advantage can also, if ungoverned, produce the kind of algorithmic targeting documented in Gaza, the kind of autonomous escalation that no treaty anticipated, and the kind of arms race that the world has not seen since the nuclear age. The international community is trying to write rules for autonomous weapons. Whether anyone will follow them is the subject of the next chapter.

Chapter 13: The Rules

On the morning of December 2, 2024, a delegate from Costa Rica sat in a plenary hall of the United Nations General Assembly in New York and watched the electronic tally board light up. The question before the General Assembly was Resolution 79/62, a text that called for a "comprehensive and inclusive approach" to the challenges posed by lethal autonomous weapons systems and mandated informal consultations among member states, civil society, the scientific community, and the weapons industry. The resolution had been negotiated for months in committee. It had been debated, amended, watered down, strengthened, watered down again. It was not a treaty. It was not a ban. It was not, in any legally binding sense, anything at all. It was an acknowledgment, the international community's admission that a category of weapon existed for which no legal definition had been agreed upon, no treaty had been signed, and no enforceable constraint had been imposed.

The vote took less than a minute. The board read: 166 in favor. 3 against. 15 abstentions.

The three votes against came from Belarus, North Korea, and Russia. The abstentions included the United States, Israel, India, and the United Kingdom, the countries whose militaries were most actively developing or deploying autonomous systems. In the gallery above the plenary floor, representatives from the Campaign to Stop Killer Robots, a global coalition of over 270 non-governmental organizations in more than 70 countries, observed the count with something between vindication and frustration. They had been advocating for a legally binding treaty since 2013. After more than a decade of stalled negotiations in Geneva, where a consensus requirement allowed any single state to block progress, the General Assembly route represented a procedural breakthrough. But they also knew what 166 votes in a General Assembly resolution were worth in practice.

Nothing. Unless someone decided to turn it into something.

The Consensus Trap

The history of autonomous weapons regulation is a history of structural paralysis.

Since 2014, the primary forum for international discussion of lethal autonomous weapons systems, commonly abbreviated LAWS, has been the Group of Governmental Experts (GGE) under the Convention on Certain Conventional Weapons (CCW), the Geneva-based framework that also governs landmines, incendiary weapons, and blinding laser weapons. The CCW was designed to address weapons that cause unnecessary suffering or have indiscriminate effects. On paper, it was the natural home for autonomous weapons governance. In practice, it became the place where autonomous weapons governance went to die.

The reason was structural: the CCW requires consensus to make decisions. Any single state party can block any proposal. For over a decade, major military powers (the United States, Russia, Israel, India, and at various points the United Kingdom) used this veto power to prevent the GGE from moving beyond "discussions" toward "negotiations." The distinction between those two words, in diplomatic language, is the difference between talking about a problem and doing something about it. The CCW has been talking about LAWS since 2014. As of March 2026, it has not begun to negotiate a single binding provision.

The pattern repeated with depressing regularity. Each year, the GGE would convene in Geneva for two sessions of approximately five days each. Delegations would present national positions. Civil society organizations would deliver statements. The chair would circulate discussion papers. And at the end of the session, the group would adopt a report noting the "productive exchange of views" and recommending that the work continue. The word "negotiate" did not appear. The word "binding" did not appear. The word "treaty" did not appear.

In September 2024, 42 states delivered a joint statement declaring that they were "ready to move ahead towards negotiations" on a legally binding instrument. It was a breakthrough in ambition. It was not a breakthrough in procedure. As long as the CCW required consensus, the 42 states that wanted to negotiate could not override the handful that did not. One third of the 128 CCW states parties were calling for action. The other two thirds were either uncommitted or actively blocking. The trap held.

The December 2024 General Assembly vote was an attempt to escape that trap, to move the question out of a consensus-bound forum and into one where a two-thirds majority could compel action. General Assembly resolutions do not create law. They create political pressure. And 166 to 3, with the major military powers confined to the abstention column rather than the opposition, was significant pressure. The resolution mandated informal consultations in 2025 and urged the CCW to "address the challenges and concerns" with a view to completing the elements of a legally binding instrument.

Whether that pressure would translate into a treaty was, as of spring 2026, the most important arms control question on Earth.

What 120 Nations Want

The answer, at least in outline, had already been drafted.

The two-tiered proposal championed by the Campaign to Stop Killer Robots and supported by over 120 states is the most developed framework for autonomous weapons governance currently in play. It is not a single document but a position that has converged over years of advocacy, national statements, and multilateral declarations, most recently endorsed by the UN Secretary-General, the International Committee of the Red Cross (ICRC), and a 30-state co-sponsoring group led by Austria.

The first tier is a prohibition. It would ban the development, production, and use of autonomous weapons systems that pose what the proposal calls "fundamental moral or legal problems." Three categories would be prohibited outright: systems that target people directly, using sensors and algorithms to identify and apply force based on target profiles or data proxies such as weight, heat, or sound; systems that operate without meaningful human control; and systems whose effects cannot be adequately predicted or controlled by human operators.

The Lavender case study from Chapter 9 is the clearest illustration of what the first tier is designed to prevent. An algorithm that flags 37,000 individuals for potential killing based on behavioral patterns, with human review lasting twenty seconds per target, is precisely the kind of system the prohibition tier would ban, not because AI is inherently wrong, but because reducing human beings to algorithmic data points and delegating lethal targeting to a machine that cannot understand the value of a human life crosses a line that the proposal's supporters believe should never be crossed.

The second tier is regulation. It would impose strict rules on autonomous weapons systems not covered by the outright ban, systems that target military objects rather than people, that operate within defined parameters, and that maintain the possibility of human oversight. The regulations would require limits on types of targets, restricting lawful use to military objects. They would impose geographical and temporal constraints, defining where the system can operate, for how long, and under what conditions. They would mandate human supervision and the physical ability to intervene, override, or deactivate the system at any time. And they would require testing, reliability verification, and review before deployment.

The framework rests on a central concept that has, over the past decade, become the organizing principle of the entire debate: **meaningful human control**.

The Three Components

Meaningful human control is not a binary switch. It is not a question of whether a human is "in the loop" or "out of the loop", the framing that Paul Scharre introduced in *Army of None* and that has shaped the policy conversation since 2018. It is, rather, a set of intersecting requirements that determine whether the humans associated with an autonomous weapon, the operators, the commanders, the engineers who built it, can genuinely be said to have controlled its lethal effects.

The concept, as developed in the GGE's rolling draft text and endorsed by the ICRC, the Campaign to Stop Killer Robots, and the majority of states participating in the negotiations, has three components.

The first is the decision-making component. Operators must understand how the autonomous system works, how it identifies targets, how it arrives at its conclusions, what parameters govern its behavior. A system that operates as a black box, whose internal logic is opaque to the humans who authorize its use, cannot be meaningfully controlled. This is not a theoretical concern. The intelligence officer who spent twenty seconds per Lavender target did not understand the algorithm's scoring methodology. He did not know how many of the 37,000 flagged individuals were flagged incorrectly. He did not have the information necessary to exercise judgment. He had a prompt and a button.

The second component is technological. The weapon system must be predictable, reliable, and interruptible. It must behave in ways its operators can anticipate. It must not produce effects that exceed the parameters its operators have set. And it must include a physical mechanism — not merely a theoretical authority — for the operator to intervene, override, or deactivate it during operation. A system that cannot be stopped once activated is not a weapon under human control. It is a weapon that has been released.

The third component is operational. The use of the system must be constrained by geographic boundaries, temporal limits, and rules of engagement that are set by humans before deployment and enforceable during operation. An autonomous weapon deployed into an area with no geographic boundary, no time limit, and no restriction on what it can engage is not meaningfully controlled, regardless of how sophisticated its target identification algorithm may be. The operational constraints are not bureaucratic technicalities. They are the spatial and temporal expression of command authority, the commander's judgment imposed on the machine before it is sent to act.

Together, these three components define meaningful human control as a condition that must be designed into the system, maintained during its use, and verifiable after the fact. A system that satisfies all three allows its operators to be held accountable for what it does. A system that fails any of the three creates the accountability gap that Human Rights Watch has called the most dangerous consequence of autonomous weapons: a gap in which no one (not the operator, not the programmer, not the manufacturer, not the commander) can be identified as responsible when something goes wrong.

The American Position

The United States occupies an unusual position in the autonomous weapons debate: the country with the most sophisticated autonomous weapons policy, and one of the countries most resistant to making that policy internationally binding.

Department of Defense Directive 3000.09, first published in 2012 and updated in 2023, is the most detailed national policy on autonomous weapons in existence. The directive defines an autonomous weapon system as one that, "once activated, can select and engage targets without further intervention by a human operator." It requires that all autonomous and semi-autonomous weapon systems "be designed to allow commanders and operators to exercise appropriate levels of human judgment over the use of force." It mandates senior leadership review and certification for fully autonomous weapons that use kinetic force against human targets. And it requires testing, training, and protocols to minimize operational failures and civilian harm.

On paper, the directive is a model policy. It acknowledges that autonomy in weapons exists on a spectrum. It requires human judgment. It mandates review and certification. If every nation adopted a policy equivalent to Directive 3000.09, the Campaign to Stop Killer Robots would not be satisfied, but the worst-case scenarios for autonomous weapons governance would be substantially mitigated.

The problems with the directive are not in its text. They are in its gaps and its enforcement.

The directive does not prohibit autonomous weapons. It does not ban weapons that independently select and engage targets. It requires that such weapons be certified as compliant with the "appropriate levels of human judgment" standard, but it does not define what "appropriate" means. The standard can be satisfied, according to Congressional Research Service analysis, as long as there is broader human input in decisions about where and how to use such weapons, without requiring a human to directly authorize individual lethal strikes. Semi-autonomous hunter-killer systems that autonomously identify and attack targets do not even require the senior leadership certification that fully autonomous lethal systems undergo.

The gap between policy and practice widened further in early 2026. In January, Secretary of Defense Pete Hegseth issued an AI strategy memo directing the department to "utilize models free from usage policy constraints" and mandating that all AI contracts include "any lawful use" language, effectively requiring technology vendors to remove their own safety restrictions. The memo's framing was explicit: "the risks of not moving fast enough outweigh the risks of imperfect alignment." Speed was the priority. Governance was, at best, a secondary consideration.

The Anthropic Test Case

The consequences of that priority surfaced almost immediately.

In February 2026, the standoff between the Pentagon and Anthropic, the AI company behind the Claude language model, became the most vivid illustration of the tension between autonomous weapons governance and operational reality. Anthropic had been awarded a $200 million prototype contract by the Pentagon in July 2025. Claude was deployed on classified networks through Palantir's Maven Smart System, where it processed intelligence data, generated targeting recommendations, and simulated battlefield scenarios. The system worked. Military commanders became dependent on it.

Anthropic's CEO, Dario Amodei, drew two red lines: Claude could not be used for mass domestic surveillance of US citizens, and it could not be used to power fully autonomous weapons. Amodei stated publicly that "frontier AI systems are simply not reliable enough to power fully autonomous weapons" and that the oversight mechanisms needed to protect civilian lives and military personnel "don't exist today."

When Anthropic refused to remove those restrictions in compliance with the Hegseth memo's "any lawful use" mandate, the Pentagon designated the company a "supply chain risk to national security." The president directed every federal agency to cease using Anthropic's technology. Pentagon spokeswoman Kingsley Wilson declared that US forces would not be "held hostage by Silicon Valley ideology."

Hours later, OpenAI announced a deal with the Pentagon to deploy its own models. The deal was negotiated in days. The company's CEO, Sam Altman, later acknowledged that it was "rushed" and "looked opportunistic and sloppy."

The Anthropic standoff revealed a structural problem that Directive 3000.09 was never designed to address: in the absence of binding law, the governance of autonomous weapons depends on voluntary compliance by both governments and technology companies. When the government decides that speed matters more than safeguards, and when a willing vendor stands ready to replace an unwilling one, the safeguards collapse. The directive requires "appropriate levels of human judgment." The AI strategy memo requires "any lawful use." These two documents, both issued by the same department, point in opposite directions. The question of which one governs when they conflict is not a legal technicality. It is the question of whether American autonomous weapons policy has a floor, a line below which no amount of operational urgency can push.

As of March 2026, the answer to that question was unclear. And the rest of the world was watching.

What the ICRC Sees

The International Committee of the Red Cross has been the most authoritative institutional voice calling for legally binding rules on autonomous weapons, and its position rests on a framework that is both older and more durable than any national policy: international humanitarian law.

The principles are familiar to anyone who has studied the laws of armed conflict. The principle of distinction requires that parties to a conflict distinguish between combatants and civilians and direct attacks only against combatants and military objectives. The principle of proportionality prohibits attacks in which the expected civilian harm is excessive relative to the anticipated military advantage. The principle of precaution requires that parties take all feasible measures to minimize civilian harm.

The ICRC's argument, articulated in a 2021 position paper and reaffirmed jointly with UN Secretary-General Antonio Guterres in October 2023, is that autonomous weapons systems pose fundamental challenges to all three principles, not because the technology is imperfect today, but because the nature of algorithmic decision-making is structurally incompatible with the case-by-case moral judgments these principles demand. Distinction requires understanding context: whether the person standing near a military target is a combatant, a civilian, a child, a journalist, a medic. Proportionality requires weighing incommensurable values: military advantage against human life, operational necessity against suffering. Precaution requires adaptive judgment: recognizing when circumstances have changed and a planned strike is no longer justified.

Machines do not exercise judgment. They execute algorithms. They can be extraordinarily good at pattern recognition, at matching a visual signature to a training dataset, at calculating a trajectory, at optimizing a strike pattern against a target set. But the moral reasoning that international humanitarian law demands of combatants is not pattern recognition. It is the distinctly human capacity to weigh competing values, to recognize the full weight of what it means to take a life, and to accept accountability for the decision.

Human Rights Watch, a co-founder of the Campaign to Stop Killer Robots, has extended this argument into the domain of human rights law. In their analysis, autonomous weapons violate the right to life because they cannot meet the legal standard that lethal force must be necessary, proportionate, and used only as a last resort. They violate human dignity because they reduce individuals to data points, what HRW calls "digital dehumanization," the process by which algorithms strip human targets of their individuality and reduce them to coordinates in a targeting system. And they create an accountability gap: because AI systems that rely on deep learning make determinations through opaque, "black box" processes, it is practically and legally impossible to punish the machine, and extremely difficult to hold operators, programmers, or commanders responsible for outcomes they could not predict or understand.

The joint appeal from Guterres and ICRC President Mirjana Spoljaric could not have been more direct: states must "launch negotiations of a new legally binding instrument to set clear prohibitions and restrictions on autonomous weapon systems and conclude such negotiations by 2026."

The clock was running.

The March 2026 Reckoning

Three months before this book went to press, the GGE on LAWS convened in Geneva for what may prove to be the most consequential session in its history.

From March 2 to 6, 2026, delegations conducted paragraph-by-paragraph readings of the rolling draft text that the GGE chair had circulated in December 2025. The draft contained elements for a possible legally binding instrument, not a treaty yet, but the skeleton of one. At the start of the week, over 40 states had expressed support for moving to formal negotiations on the basis of this text. By the end of the week, that number had grown to over 70, after a group of African states joined the call. A majority of the GGE's members were now on record: they wanted to negotiate.

The United States did not.

During the debate on the draft text's provisions regarding human control, the central concept that the majority of states, the ICRC, and the UN Secretary-General had identified as the non-negotiable core of any instrument, the US delegation refused to accept the term "human control." It proposed instead the phrase "good faith human judgement and care."

The distinction sounds technical. It is not. "Human control" implies a structural requirement, a system designed so that humans can direct, intervene in, and override its operations. "Good faith human judgement and care" implies a subjective standard, a commander's sincere belief that appropriate thought went into the decision. The first can be verified. The second cannot. The first creates accountability. The second creates plausible deniability.

Many delegations pushed back. The vast majority of states have maintained, consistently, for over a decade, that human control over weapons is imperative. They would not accept weaker language that, in the view of arms control advocates, was designed to preserve the operational flexibility of the world's most advanced military rather than to protect the civilians that international humanitarian law exists to safeguard.

The context for the US position was impossible to ignore. At the same moment the American delegation was proposing "good faith human judgement and care" as a substitute for "human control," the American military was reportedly using Claude to generate targeting recommendations in an active combat operation in Iran, a system whose developer had been blacklisted for insisting that it not be used for fully autonomous weapons. The gap between what the United States said in Geneva and what it did in the field was not subtle.

The Author's Position

This book has spent twelve chapters documenting a technology revolution. It is time to take a position on how that revolution should be governed.

A complete ban on autonomous weapons is neither achievable nor desirable. The technology is too widespread. Too many countries, too many companies, and too many armed groups already possess or are developing autonomous capabilities. The components of an autonomous weapon, commercial drones, open-source software, 3D-printed airframes, are available on the open market. A ban that cannot be verified, that cannot be enforced, and that the major military powers will not sign is not governance. It is aspiration.

But the absence of a ban does not mean the absence of rules. The path forward is a legally binding international framework centered on meaningful human control, the three-component standard described above, with specific, verifiable prohibitions and regulations.

The prohibitions should be narrow and defensible. Systems designed to target people through algorithmic profiling (the Lavender model, in which an algorithm generates a kill list of individuals based on data patterns) should be prohibited outright. Not because algorithmic identification has no military utility, but because reducing human beings to probability scores in a targeting system, with twenty seconds of human review and pre-authorized civilian casualty thresholds, is incompatible with the principles of distinction, proportionality, and human dignity that international humanitarian law exists to protect. Systems that cannot be deactivated, overridden, or interrupted by a human operator during use should also be prohibited. A weapon that cannot be stopped is not a weapon under control.

The regulations should be broad and enforceable. Every autonomous weapon system not covered by the prohibitions should be required to maintain the three components of meaningful human control: the decision-making component (operators must understand how the system works), the technological component (the system must be predictable, reliable, and interruptible), and the operational component (geographic, temporal, and targeting constraints set by humans before deployment). Compliance should be verifiable through mandatory testing, reporting, and review, not through self-certification by the deploying state.

Why should the United States support such a framework, given that it would constrain American military flexibility?

Because the framework constrains everyone else too. The United States already has Directive 3000.09, a policy that, when followed, approximates meaningful human control. The countries that do not have such policies, Russia, China, Iran, and the dozens of nations and non-state actors acquiring autonomous systems, are the countries for whom a binding framework would impose real constraints. A treaty that codifies what the United States already claims to do, while requiring adversaries to accept verifiable limits they currently reject, is not a concession. It is a strategic advantage.

And because the alternative, a world in which autonomous weapons proliferate without binding legal constraints, in which any algorithm can generate a kill list and any commander can rubber-stamp it, in which the accountability gap grows wider with every deployment, is a world in which the principles the United States claims to defend become inoperable. The rules of war exist because some things should not be done even by the side that is winning. If those rules do not apply to autonomous weapons, they will eventually not apply to anything at all.

The Clock

The GGE's mandate runs through one more session, scheduled for August 31 to September 4, 2026. After that, the CCW Seventh Review Conference convenes in November 2026, the moment when states must decide whether to formally launch treaty negotiations.

The UN Secretary-General has called for conclusion of those negotiations by the end of 2026. The ICRC has endorsed the same timeline. Over 70 states have called for negotiations to begin on the basis of the rolling text. The political will, measured in state count, is overwhelming.

The structural obstacles remain formidable. The CCW's consensus requirement means that any single state can block the launch of negotiations at the Review Conference. The United States, Russia, and India have all declined to support binding measures. The semantic battle over "human control" versus "good faith human judgement and care" (fought word by word, clause by clause, in the airless conference rooms of the Palais des Nations) reflects a deeper disagreement about whether autonomous weapons should be governed by verifiable standards or by national discretion.

The clock that the ICRC and the Secretary-General set, conclude negotiations by the end of 2026, was never realistic as a timeline for a completed treaty. Arms control negotiations, even when they enjoy broad support, take years. The Chemical Weapons Convention took six years from mandate to signing. The Ottawa Treaty on landmines took two years, and it was considered miraculously fast. A treaty on autonomous weapons, touching on the most advanced military technologies of the most powerful nations, negotiated in an era of great-power competition and institutional distrust, will not be completed in months.

But the clock is not about the treaty. The clock is about the technology. Every month that passes without binding rules, autonomous weapons systems are deployed in new conflicts, integrated into new military architectures, and sold to new customers. The Lavender model is studied and replicated. The kill chain compresses further. The Gap narrows. The systems that a treaty might constrain become more entrenched, more distributed, and harder to control. The arms control community learned this lesson with nuclear weapons: the time to write the rules is before the arsenals are built, not after.

The rules matter. But rules without a framework for implementation are aspirations, not governance. The final chapter synthesizes everything this book has documented (the technology, the stakes, the competition, the ethics) into a single prescription: the Hunter-Killer framework applied to the most important question of all. Not whether autonomous weapons will exist. They already do. But whether humans will remain in meaningful control of the decision to take a life.

Chapter 14: Hunter, Killer, Human

On a Tuesday afternoon in March 2026, Adam Bry sat in a glass-walled conference room at Skydio's headquarters in San Mateo, California, and considered a question he had been circling for a decade. Outside, the parking lot held the same electric vehicles and the same understated signage that mark every defense-tech company that would prefer not to be noticed from the street. Inside, the office looked like what it was: the command center of the largest drone manufacturer in the United States, a company that had just received the largest single-vendor small drone order in Army history, whose products were deployed by every US military branch and 29 allied nations, and whose CEO had cofounded the company in a university robotics lab twelve years earlier because he believed that autonomous flight would change the world.

I asked Bry the question that had animated this entire book: as AI compresses the gap between sensing a target and striking it, between hunter and killer, where does the human lose meaningful control?

He paused for a long time before answering. When he spoke, he described a philosophy that was both simple and radical. "The drone should work for the person, not the other way around," he said. "One person should have five, 10, 50 drones doing useful work on their behalf. And the only way you're going to get there is with AI and autonomy, making the thing smart enough to fly itself."

This was the standard Bry pitch, the one he had given to investors, military officials, and conference audiences for years. But then he added something else, and his voice shifted. He was no longer selling a product. He was describing a future he had thought about carefully and was not entirely comfortable with.

"We are on a faster track to artificial general intelligence than much of our civilization realizes," Bry said. "But even in a world with AGI, human intent, initiative, empathy, and judgment will remain essential ingredients for ethical conflict. In the optimistic view, future conflicts will have fewer human casualties and be direct technology and production competitions."

I wrote down the phrase: *essential ingredients for ethical conflict.* Coming from the CEO of an autonomous drone company — a man who had spent a decade building machines that navigate and sense without human input — this was not a throwaway line. It was a design philosophy stated as a moral position. The drone handles the mechanics. The human provides the judgment. The AI frees the operator from the cognitive burden of flight so that the operator can focus on the part of the job that cannot be automated: deciding what to do with the information the machine has gathered.

This is the optimistic view. It is also not the only plausible one.

Two Futures

The evidence assembled in the preceding thirteen chapters supports two scenarios. Both are plausible. Both are already partially underway. Which one prevails depends on choices that governments, militaries, companies, and engineers will make in the next few years.

In the first scenario, call it the Bry scenario, autonomous systems extend human capability without replacing human judgment. One operator manages 50 drones. Each drone flies itself, navigates contested airspace, builds a three-dimensional map of the battlefield, identifies objects of interest, and presents its findings to the human. The human decides what matters. The human assigns priorities. The human authorizes action. The AI is fast. The human is responsible. Conflicts become contests of industrial capacity and technological sophistication, with fewer soldiers in harm's way and more precise application of force. The Gap between hunter and killer narrows on the sensing side (machines see more, faster, with greater persistence) but the killing side retains a human checkpoint. The drone works for the person.

In the second scenario, call it the Lavender scenario, the compression of the Hunter-Killer gap does not stop at sensing. Algorithms generate kill lists. Humans rubber-stamp them. Pre-authorized civilian casualty thresholds replace case-by-case proportionality judgments. The twenty-second review becomes the ten-second review becomes no review at all, because the speed of autonomous engagement exceeds the speed of human cognition. Electronic warfare severs the communications link, and the drone completes the strike on its own, not as a fallback, but as the design. The accountability gap widens. The principles of distinction, proportionality, and precaution become engineering specifications rather than moral obligations. Wars become cheaper to start, because the political cost of casualties drops when the casualties are borne by the other side's civilians rather than by your own soldiers. The international arms control community warned about this future for over a decade. As of early 2026, it is no longer a warning. It is a description.

The Lavender scenario does not require malice. It does not require a commander who wants to kill civilians. It requires only the logic of operational tempo: there are too many targets, the threat is too urgent, and the algorithm is too fast for human review to keep pace. The incentive structure points in one direction. When the machine offers speed and the human offers delay, institutions under pressure will choose speed. That is not a moral failing. It is an organizational inevitability, unless the system is designed, from the ground up, to prevent it.

The Bry scenario and the Lavender scenario are not theoretical. They are both happening simultaneously, in different places, under different commands, with different consequences. The Skydio X10D flying ISR missions in a GPS-denied environment, with a human operator deciding what to do with its sensor data, is the Bry scenario in practice. The Lavender algorithm flagging 37,000 individuals for potential killing, with human review measured in seconds, is the Lavender scenario in practice. They are both on the Hunter-Killer spectrum. They represent different answers to the same question: where does the human retain meaningful control?

The honest assessment is that both scenarios are likely to coexist for years, deployed by different countries with different values and different accountability structures. The question for the United States is not which scenario will prevail globally, that is beyond any single nation's control. The question is which scenario the United States chooses for itself, and whether it builds institutions and norms that push the global trajectory toward the first scenario rather than the second.

This chapter argues that there is a third option, one that takes the best of both scenarios and rejects the worst. It is not original to this book. Its intellectual foundation was laid by Paul Scharre in *Army of None*. Its operational principles are being demonstrated by the companies described in Chapter 12. Its moral logic follows directly from the framework this book has been building since the Introduction.

The Centaur

In the mid-1990s, world chess champion Garry Kasparov lost a match to IBM's Deep Blue and experienced what he later described as a revelation. The machine had beaten the best human player on Earth. But Kasparov observed something that the headlines missed: when a human and a computer worked together, what he called "centaur chess", the team consistently outperformed either a standalone human grandmaster or a standalone supercomputer. The human provided strategy, intuition, and the ability to recognize when the computer's recommendations did not make sense. The computer provided speed, calculation depth, and the ability to evaluate millions of positions per second. Together, they were better than either alone.

Paul Scharre borrowed the centaur metaphor for warfare in *Army of None* and proposed the **centaur warfighter** as a policy model for autonomous weapons. The centaur model divides tasks by what each partner does best. The machine handles the functions that benefit from speed, scale, and computational power: processing sensor data, navigating terrain, identifying potential targets, calculating trajectories, coordinating with other autonomous systems, and presenting options to the human operator. The human handles the functions that require moral reasoning, contextual judgment, and accountability: deciding whether to engage, assessing proportionality, interpreting ambiguous situations, recognizing when the machine's recommendation is wrong, and accepting responsibility for the outcome.

The centaur model is not human-in-the-loop in the traditional sense, where a human must press a button to authorize every individual engagement. In a swarm of 50 drones operating in contested airspace, the operator cannot individually authorize every movement, every sensor sweep, every navigation decision. The mechanical tasks must be autonomous. The centaur model does not require human authorization for every action. It requires human authority over every lethal decision. The human does not fly the drone. The human does not aim the camera. The human decides whether someone dies.

The distinction matters because the most common objection to maintaining human control is that it is impractical at the speed of modern warfare. The objection assumes that human-in-the-loop means human authorization of every mechanical function. It does not. The centaur model separates the tasks that benefit from machine speed (navigation, tracking, coordination, data processing) from the task that demands human judgment: the authorization of lethal force. A swarm of 50 autonomous drones can navigate, sense, and present targeting options at machine speed while the human operator retains the authority to approve, modify, or reject each engagement recommendation. The bottleneck is not the human. The bottleneck is the absence of systems designed to present lethal decisions in a format that humans can evaluate quickly and responsibly. Building those interfaces, decision-support systems that compress information without compressing accountability, is an engineering challenge, not a philosophical concession.

This is where the Hunter-Killer framework becomes not just an analytical tool but a moral compass.

The Spectrum as Compass

Every system documented in this book exists somewhere on the Hunter-Killer spectrum, the continuum between sensing a target and striking it. The central question is not where a system falls on that spectrum. It is where on that spectrum the human loses meaningful control.

Consider three cases, drawn from the preceding chapters.

The first is the hunter-dominant model: Skydio's X10D. The drone is an autonomous sensor platform. It flies itself, navigates without GPS, builds three-dimensional maps, and identifies objects of interest using nine deep neural networks running simultaneously on an NVIDIA Jetson Orin GPU. It does not carry weapons. The human operator receives the drone's sensor data and decides what to do with it. The autonomy is concentrated entirely on the sensing side of the spectrum. Human control over lethal action is not compressed, it is not even engaged, because the X10D does not kill. This is the hunter end of the spectrum, and it is relatively uncontroversial. Autonomous ISR should be encouraged. More autonomous sensing means better information for human decision-makers. The risk is not that the X10D will make a lethal decision without human input. The risk, if there is one, is that the information it provides will be fed into a system further down the spectrum that does.

The second is the killer-dominant model: the Lavender system. The algorithm processes mass surveillance data and assigns each person in its database a probabilistic score indicating the likelihood that they are a militant. The system generates a list of 37,000 names. Human review lasts twenty seconds per target. The human contribution to the targeting decision is, by the operator's own testimony, "zero added-value." The autonomy has crossed the entire spectrum (from sensing to targeting to engagement authorization) with the human reduced to a rubber stamp. This is the killer end of the spectrum. It is the case in which meaningful human control has been lost. Not because the technology failed, but because the humans who deployed it chose to treat algorithmic output as sufficient justification for killing.

The third is the centaur model: the model this book advocates. AI handles the speed-of-light functions, sensing, tracking, navigating, coordinating, presenting options. Humans handle the moral-weight functions, deciding whether to engage, assessing proportionality, accepting accountability. The division is not arbitrary. It follows from what machines do well and what humans do well. Machines process data at scale. Humans exercise judgment under uncertainty. The centaur model does not ask humans to do what machines do better, or machines to do what humans do better. It assigns each task to the partner best suited to perform it, and reserves the irreducible moral act of deciding to take a life for the partner capable of understanding what that decision means.

The Gap (the space between hunter and killer that this book has tracked from its first page) is where that moral act lives. As AI compresses the Gap, the space for human judgment shrinks. But the Gap does not have to close. It can be defended. It can be designed into weapons systems, codified in policy, and enforced through law. The centaur model is the architectural expression of that defense: a weapons architecture that compresses the Gap on the sensing side, where compression makes the system more capable, and preserves it on the killing side, where compression makes the system less accountable.

The Gap is not a technical specification. It is a moral compass. The closer a system moves toward autonomous killing, the more urgently it demands human oversight. The farther it sits on the hunter end of the spectrum, the more freely it can operate. The framework applies to every system in this book, and to every system that will be built after it.

Five Prescriptions

This book has documented a revolution. It has traced the technology from Ukrainian basements to Silicon Valley garages to Chinese state factories. It has followed the money, the patents, the contracts, and the casualties. It has examined the competition between the United States and China, the proliferation of lethal autonomy to non-state actors, the ethical catastrophe of algorithmic targeting in Gaza, the structural paralysis of international arms control, and the emergence of an American drone industrial base that did not exist a decade ago. What follows are five specific, actionable, debatable prescriptions. They are designed to be cited in policy discussions, challenged in think tank papers, tested against competing proposals, and, most importantly, implemented.

Prescription One: Build the base. The United States must build a domestic drone industrial base capable of producing autonomous systems at scale. The foundation already exists: Skydio for autonomous sensing, Anduril for networked command and lethal systems, Shield AI for autonomous piloting. But the current production capacity, thousands of drones per year, is orders of magnitude below what the threat demands. Ukraine produced over 4 million drones in 2024. China's production goal is measured in millions. The United States produces thousands. Closing that gap requires industrial policy: long-term production contracts that give manufacturers the demand signal to invest in capacity, domestic supply chain development that eliminates dependence on Chinese components, and manufacturing facilities, like Anduril's Arsenal-1 in Ohio, that are designed for volume from the ground up. The Skydio battery crisis of 2024, when China's sanctions against Dongguan Poweramp nearly shut down America's most important tactical drone manufacturer, demonstrated that a defense-tech company with American technology and American engineers can still be killed by a Chinese supply chain. The drone industrial base is not a nice-to-have. It is a strategic necessity equivalent to the shipbuilding capacity that determined the outcome of the Second World War.

Prescription Two: Buy at speed. The Pentagon must adopt acquisition timelines measured in hours and days, not years. The 72-hour Skydio order should be the standard, not the exception. This requires expanding Other Transaction Authority pathways, reducing the compliance burden on companies that have already been cleared for defense work, and institutionalizing the Barrier Removal Board and CTO Action Group established in the 2026 AI strategy. But it also requires something harder than procedural reform: cultural change within the acquisition workforce. The Pentagon has been trying to reform acquisition since the Packard

Commission in 1986. Forty years of reform efforts have produced incremental improvements and no fundamental change. The people who manage defense procurement are trained to minimize risk, ensure fairness, and follow process. Those values are not wrong. They are insufficient for wartime speed. The Replicator initiative, launched with the explicit goal of fielding thousands of autonomous systems within two years, fielding only hundreds by its deadline, is the definitive case study. The drones existed. The factories existed. The procurement system could not connect supply with demand. The United States does not have a technology problem. It has a buying problem. Until the institution that buys weapons can move as fast as the institutions that build them, the drone revolution will remain a revolution that the United States watches rather than leads.

Prescription Three: Support the treaty. The United States must support a legally binding international framework for autonomous weapons centered on meaningful human control. Not because it constrains American advantage (the framework would codify what the United States already claims to do under Directive 3000.09) but because it constrains adversaries and establishes norms. The two-tiered proposal supported by over 120 states is a reasonable starting point: prohibitions on systems that target people through algorithmic profiling or operate without meaningful human control, and regulations for all other autonomous systems including testing, human oversight, and operational constraints. The United States has historically benefited from the international legal order more than any other nation. Undermining that order by blocking autonomous weapons governance does not make America safer. It makes the world more dangerous. The semantic maneuver of substituting "good faith human judgement and care" for "human control" at the March 2026 GGE was not a win for American interests. It was an abdication of American leadership on the most important arms control challenge since nuclear weapons.

Prescription Four: Keep the human in the kill decision. Every autonomous weapons system deployed by the US military must maintain a human decision point for lethal action. Not a rubber stamp. Not a twenty-second review. A genuine decision by a human who understands the system's recommendation, has the authority and the time to override it, and will be held accountable for the outcome. This is the centaur model, not the Lavender model. It means designing autonomous weapons systems so that the AI presents options and the human chooses, with enough context, enough time, and enough authority to make the decision meaningful. It means setting minimum review standards based on the complexity of the targeting decision, not on the volume of targets to be processed. It means building decision-support interfaces that compress information without compressing accountability, showing the operator what the AI has found, how confident it is, what the risks are, and what the alternatives are, in a format that allows a genuine judgment, not a perfunctory click.

It also means rejecting the argument that combat tempo requires the human to be removed from lethal decisions. The practical reality that in the fastest engagements a human cannot respond within the engagement window is not an argument for removing the human. It is an argument for designing systems that do not require engagement within that window, systems that can decline to engage when the human cannot be consulted, rather than systems that engage because the human was not fast enough to stop them. The willingness to accept a missed engagement, to let a target escape because the human could not authorize the strike in time, is the price of accountability. It is a price worth paying. The alternative, a system that strikes without authorization because the window was too short for human review, is the Lavender model by another name.

Prescription Five: Fund the counter. The United States must invest in counter-drone systems as aggressively as it invests in drone systems. The same proliferation that empowers American forces and American allies also empowers adversaries, terrorist organizations, and any individual with $500 and access to commercial drone components. The Houthi drone and missile campaign against commercial shipping in the Red Sea demonstrated that non-state actors with inexpensive autonomous weapons can impose strategic costs on the world's most powerful navies. The FPV drones destroying main battle tanks in Ukraine demonstrated that $200 weapons can negate $10 million platforms. The same economic logic that makes drones revolutionary for offense makes them existentially threatening for defense.

The $7.5 billion authorized for counter-unmanned aircraft systems in the FY2026 NDAA is a start. It is not enough. Counter-drone defense must become a core military competency, integrated into every unit's training, every base's defenses, and every operational plan's threat assessment. Anduril's $20 billion Army counter-drone contract, built around the Lattice platform, represents the right architectural approach: a software-defined mesh that integrates sensors, interceptors, and electronic warfare into a coordinated defense. But counter-drone defense cannot remain a single program or a single company's responsibility. It must be as pervasive as air defense was in the Cold War, embedded at every echelon, from individual platoons to combatant commands. The drone revolution is not unidirectional.

The country that builds the best drones and the best defenses against drones occupies a fundamentally different strategic position than the country that builds only the best drones. Offense without defense is a capability. Offense with defense is a strategy.

These five prescriptions are not a complete policy. They are a framework for action, specific enough to implement, broad enough to debate, and grounded in the evidence documented in the preceding thirteen chapters. They will be challenged. They should be. The mark of a serious prescription is not that everyone agrees with it. It is that disagreement is productive, that the arguments against lead to better alternatives rather than comfortable inaction.

The Irreducible Act

This book began in a basement in Zaporizhzhia, with a drone operator guiding a $200 weapon toward a $3 million tank. It has traveled from Ukrainian front lines to Silicon Valley boardrooms, from Chinese drone factories to the conference rooms of the United Nations. It has followed a single question across fourteen chapters: as AI compresses the gap between sensing a target and striking it, between hunter and killer, what happens to human judgment?

The answer is not predetermined. The answer is a choice.

The technology that Adam Bry is building at Skydio (autonomous sensing that frees humans from the mechanics of flight so they can focus on the decisions that matter) is one choice. The system that generated a list of 37,000 names and invited a human to rubber-stamp each one in twenty seconds is another. Both use AI. Both operate on the Hunter-Killer spectrum. Both represent a version of the future that is already arriving.

The centaur model (AI speed married to human judgment, machine capability yoked to human accountability) is the choice this book advocates. Not because it is the easiest option. The easiest option is full autonomy: let the algorithm handle everything, from target identification to engagement, and accept whatever the machine decides. Full autonomy is faster, cheaper, and more scalable. It is also the option that removes the moral agent from the moral act.

The harder option is to build systems that are fast enough to compete with adversaries who have no interest in human oversight, and governed enough to be trusted with the most consequential decision a state can make, the decision to use lethal force. Speed without governance is recklessness. Governance without speed is irrelevance. The centaur model insists on both.

Adam Bry acknowledged this tension in the interview in San Mateo. He recognized that the track toward artificial general intelligence is faster than most people realize. He also insisted that even in a world with AGI, a world where machines can match or exceed human cognitive capabilities across every domain, "human intent, initiative, empathy, and judgment will remain essential ingredients for ethical conflict."

I pressed him on this. Why? If AGI can process more information, assess more variables, and calculate more precisely than any human, why should the human remain in the decision loop?

Because the decision to take a human life is not an optimization problem. It is not a question that can be resolved by processing more data or running a more sophisticated algorithm. It is a moral act, an assertion that the situation is grave enough, the threat is imminent enough, and the alternatives are insufficient enough to justify ending a life. That assertion requires something that no algorithm possesses: an understanding of what a life is worth. Not a probability score. Not a data point. An understanding.

The Gap between hunter and killer is where that understanding lives. It is the space in which a human being, fallible, slow, burdened by conscience, looks at what the machine has found and decides what to do about it. The machine can compress that space. AI can make it faster to sense, faster to track, faster to present options, faster to coordinate. But the irreducible act, the decision, must remain human. Not because humans are better at it in some measurable, quantifiable way. But because a decision to take a life that no human made is a decision for which no human is accountable. And a world in which lethal force is applied without accountability is a world in which the laws of war, the principles of distinction and proportionality, and the fundamental commitment to human dignity that underlies the entire framework of international humanitarian law become meaningless.

These prescriptions are designed to defend that space. Build the industrial base, so the United States does not cede autonomous capability to adversaries. Buy at speed, so the technology reaches the warfighter before the threat arrives. Support the treaty, so the rules apply to everyone. Keep the human in the kill decision, so accountability survives the compression. Fund the counter, so the drones that adversaries build face the same defenses we build for ours.

The Gap is not a weakness. It is the space where ethics lives. Defend it.

The prescriptions are clear. The evidence is documented. The framework is established. What remains is the question that opened this book: what happens next? The Conclusion returns to where the book began, and looks forward to the war that has not yet been fought.

Conclusion: The Next War

The operator sat in the same basement in Zaporizhzhia, two floors below the same apartment block that had lost its upper stories to a Russian glide bomb, though the rubble had been cleared now, the rebar cut and stacked, the surviving floors weatherproofed with sheets of corrugated metal bolted to the exposed concrete. He was twenty-eight years old, callsign still "Lens," and he had been doing this for two years and four months. The room smelled the same, solder flux, stale coffee, the ozone tang of lithium-polymer batteries. But almost everything else had changed.

The FPV goggles were gone. The single-stick controller was gone. In their place sat a ruggedized tablet (a Ukrainian-manufactured unit, built by a Kyiv company that had not existed when the full-scale invasion began) running an interface that Lens's commanders referred to as a Swarmer-type platform. On the screen, twenty blue icons moved in a loose formation over the same gray zone east of Orikhiv where Lens had once guided a single drone into a single tank. Each icon represented an FPV strike drone equipped with a machine-vision targeting module and a fragmentary munition. The drones navigated without GPS (the Russians had jammed that out of existence over most of the front more than a year ago) using onboard visual-inertial odometry to hold their positions relative to terrain features and to each other. Lens did not fly them. He did not see through their cameras. He set a target area on the map, defined engagement parameters (vehicle types authorized for strike, minimum confidence threshold, distance from known civilian structures) and monitored the swarm as it executed.

Twenty drones. One operator. Two years earlier, that ratio had been one to one.

The Gap had compressed again. It had been compressing since the first chapter of this book, since the first FPV drone dove into the first tank on a thermal feed, guided by a human thumb on a stick. Now the thumb was off the stick. The AI handled navigation, target identification, attack coordination, and battle damage assessment. The system found, fixed, tracked, and struck. Lens's role had shifted from pilot to something closer to a commander: he told the swarm what to do, set the boundaries of acceptable action, and watched as twenty semi-autonomous aircraft executed within those boundaries. He could override any individual drone. He could abort the entire mission with a single input. He could intervene at any point between detection and detonation.

But the interval in which he could intervene had shrunk from minutes to seconds. And the number of simultaneous engagements he was expected to supervise had grown from one to twenty.

The Compression Is Not Theoretical

This is what the autonomous weapons revolution looks like in practice, not the science-fiction scenario of a fully autonomous robot deciding to kill without any human involvement, but something more mundane and in many ways more urgent. A human is still in the loop. A human still authorizes. A human still bears accountability. But the nature of that human role has changed. The operator is no longer the one who finds the target, tracks it, confirms it, and guides the weapon onto it. The operator is the one who defines the rules under which a machine does all of those things, and who watches, across twenty simultaneous feeds compressed into a single interface, for the moment when the machine gets it wrong.

The question this book has spent fourteen chapters assembling the evidence to answer is whether that constitutes meaningful human control. Whether a person supervising twenty autonomous engagements simultaneously can exercise the kind of judgment, the contextual, moral, human judgment that distinguishes a combatant from a civilian, a legitimate target from a protected site, a proportionate strike from a war crime, that the laws of armed conflict require and that basic decency demands.

The honest answer: it depends on the system, the operator, the tempo, and the stakes. And that answer is precisely why the Hunter-Killer framework matters.

The Framework as a Tool

Every system in this book, from the $200 FPV drone in Lens's hands two years ago to the $60 billion Lattice platform in Anduril's server racks, from Swarmer's Styx coordinating over 100,000 combat missions to the Israeli military's Lavender flagging 37,000 suspected militants, from Baykar's K2 swarm to China's 200-drone demonstrations, exists on the spectrum between hunter and killer. The hunter side senses, finds, identifies. The killer side acts, strikes, destroys. Between them lies the Gap: the space where a human being once made a decision, and where human judgment either persists or has been engineered away.

The framework is not academic. It is operational. It applies to the next drone system you read about, the next defense contract you evaluate, the next policy debate you enter, the next piece of combat footage that crosses your screen. For any autonomous weapon, three questions locate it on the spectrum. Where does the system fall between hunting and killing (does it find targets, strike them, or do both? Where on that continuum does the human retain meaningful control) can the operator understand, intervene in, and accept accountability for the system's actions? And is the Gap between sensing and striking wide enough for human judgment to survive, or has the Compression reduced the human's role to a rubber stamp?

Those three questions are the tool this book leaves you. They work in a Pentagon briefing. They work in a Senate hearing. They work in a venture capital due diligence call. They work in a conversation with someone who just watched drone footage from Ukraine or Gaza and asked the question that matters: *What is happening?*

Five Prescriptions, One Warning

The evidence assembled across fourteen chapters supports five specific positions. They are restated here not as summaries but as imperatives, because the window for acting on them is open but narrowing.

First: the United States must build a domestic drone industrial base capable of producing autonomous systems at scale. Skydio, Anduril, and Shield AI are the foundational companies, but the base must be broader, deeper, and faster than any single firm. Ukraine built its production capacity from nearly nothing to 4 million drones per year in under three years. The United States can match that trajectory if it treats drone manufacturing as a strategic priority on par with semiconductor fabrication, and it must, because the alternative is dependence on foreign supply chains that adversaries can disrupt.

Second: the Pentagon must compress its acquisition timelines from years to days. The 72-hour Skydio procurement (2,500 X10Ds, $52 million, from bid to award before the weekend was over) demonstrated that rapid acquisition is not a fantasy. It is a choice. The Replicator initiative's failure to field at its original scale was not a technology problem. It was an institutional one. The technology exists. The procurement system does not match it. Fix the system or lose the race.

Third: the United States must support a legally binding international framework for autonomous weapons centered on meaningful human control. Not because a treaty constrains American advantage (the US already maintains human decision points in its kill chain) but because a treaty constrains adversaries who do not, establishes norms that allies can enforce, and provides the legal architecture to hold accountable those who deploy fully autonomous lethal systems without human oversight. The UN General Assembly's 166-nation vote was a starting point, not a finish line.

Fourth: every autonomous weapons system deployed by the US military must maintain a human decision point for lethal action. The centaur model, AI handles speed, humans handle moral weight, is not merely an ethical aspiration. It is a practical doctrine that produces better outcomes than full autonomy in complex, ambiguous environments where the cost of a misidentification is measured in civilian lives. The Lavender model, where a human rubber-stamps algorithmic kill lists in an average of twenty seconds per target, is not meaningful human control. It is the performance of control without the substance. The United States must choose the centaur over the rubber stamp, and must build its systems to make that choice possible, not merely permissible.

Fifth: the United States must invest in counter-drone systems as aggressively as it invests in drone systems. The $20 billion Anduril C-UAS contract is a beginning, not an end. The same proliferation dynamics that put autonomous strike capability in the hands of allies also put it in the hands of adversaries, non-state actors, and anyone with $200 and access to a Telegram channel. Defense against swarms is as strategically important as the swarms themselves, and far less glamorous, which is why it has been chronically underfunded. That must change.

The Decision That Remains

Lens powered down the tablet. The swarm's last drone had returned to the recovery point, a cleared patch of farmland three kilometers behind the front, where a two-person team collected the surviving aircraft, swapped batteries, and reloaded munitions for the next cycle. Of the twenty drones he had dispatched, fourteen had struck targets. Three had been brought down by Russian electronic warfare. Two had returned after their confidence threshold was not met, the system's targeting algorithm could not classify the objects in the engagement zone with sufficient certainty, and the rules Lens had set required a minimum probability before authorizing a strike. One had lost its navigation lock and auto-returned to the recovery point rather than flying blind.

Twenty drones. One human. Fourteen engagements in forty minutes. Two years ago, those fourteen strikes would have required fourteen operators, fourteen sets of goggles, fourteen individual decisions stretched across fourteen separate moments of human attention. The Compression had delivered exactly what it promised: more lethality, fewer operators, faster execution.

And yet.

Lens had intervened twice during the mission. Once, he had overridden a targeting solution when the system flagged a vehicle near a structure he recognized from an earlier reconnaissance pass as a field hospital. The algorithm had classified the vehicle correctly, it was, in fact, a military logistics truck. But its proximity to the hospital made the strike disproportionate under the engagement rules Lens carried in his head, rules no targeting algorithm had been taught, rules that lived in the judgment of a 28-year-old man sitting in a basement in Zaporizhzhia. The second time, he had paused the swarm entirely for ninety seconds while he confirmed with his battalion that a cluster of heat signatures on the western edge of the target area were not friendly forces conducting a patrol that had not been logged in the shared operational picture.

Two interventions. Forty minutes. Twenty drones. The Gap was narrow, narrower than it had ever been, narrower than most policymakers in Washington or Geneva or New York understood. But it existed. A person had looked at what the machine was about to do and said *no.* A person had stopped the system and asked a question the system could not ask itself. A person had applied judgment that no algorithm, however sophisticated, possessed: the judgment born of context, experience, moral weight, and the irreducible human understanding that the thermal bloom on a screen was not a data point but a life.

The Gap held. For now.

Lens climbed the stairs and stepped outside. The night air was cold, late March in southern Ukraine, the ground still frozen beneath a thin mud layer that would not thaw for another month. Somewhere to the east, artillery flickered along the horizon, the sound arriving seconds after the light. Somewhere above, drones flew, his, theirs, autonomous and piloted, hunting and killing in patterns no single human could track. The war continued. It had continued through every technological revolution described in this book, through FPV swarms and AI targeting and mesh networks and electronic warfare and counter-drone systems and $20 billion contracts and 166-nation votes. The machines had gotten faster, smarter, cheaper, more lethal, more autonomous. The war had not ended. The dying had not stopped.

But the decision — the human decision, the irreducible moral act of choosing whether another person lives or dies — remained in human hands. Not because the technology required it. The technology was ready, today, to remove the human entirely. It remained in human hands because people, operators, commanders, engineers, policymakers, citizens, had chosen to keep it there. Had fought to keep it there. Had built systems and written laws and enforced norms to keep it there.

Whether it stays there depends on what happens next.

A Quick Favor

If you enjoyed this book, please consider leaving a short review on Amazon or wherever you purchased it. Even a sentence or two helps other readers decide if this story is for them — and it's the single most effective way to support an independent author.

Thank you for reading.

Selected Bibliography

Introduction and Chapter 1

Grand View Research. "Military Drone Market Size, Share & Trends Analysis Report." 2025.

Abraham, Yuval. "'Lavender': The AI Machine Directing Israel's Bombing Spree in Gaza." *+972 Magazine*, April 3, 2024.

Newdick, Thomas, and Tyler Rogoway. "Automated Terminal Attack Capability Appears To Be Making Its Way Into Ukraine's FPV Drones." *The War Zone*, March 21, 2024.

O'Grady, Siobhan, et al. "Ukraine scrambles to overcome Russia's edge in fiber-optic drones." *The Washington Post*, May 23, 2025.

Bort, Julie. "Anduril Raises $2.5B at $30.5B Valuation Led by Founders Fund." *TechCrunch*, June 5, 2025.

Clark, Kate, and Becky Peterson. "Thrive Capital, A16Z to Lead Anduril Investment at $60 Billion Valuation." *The Wall Street Journal*, March 3, 2026.

Chapters 2 and 3

Axe, David. "Russia's Electronic-Warfare Troops Knocked Out 90 Percent Of Ukraine's Drones." *Forbes*, December 24, 2022.

Malyasov, Dylan. "Ukraine's Tiny Drone Blows Up Russian Tor Air Defense System." *Defense Blog*, November 9, 2023.

Nikolov, Boyko. "Ukrainian FPVs Killed Their Most Expensive Russian Target — TOR SAM." *Bulgarian Military*, November 8, 2023.

Chapters 4 and 5

Georgetown University Center for Security and Emerging Technology. "PLA AI-Related Contract Award Notices Analysis." 2024.

Ignatius, David. "A Former Navy SEAL Saw the Future of Warfare in a Tragedy." *Washington Post*.

Scharre, Paul. *Army of None: Autonomous Weapons and the Future of War*. W.W. Norton, 2018.

Chapters 6 and 7

Brose, Christian. *The Kill Chain: Defending America in the Future of High-Tech Warfare*. Hachette Books, 2020.

Tillipman, Jessica. "Regulation by Contract: The Pentagon's AI Governance Gap." *Lawfare*, March 2026.

Chapters 8 and 9

McKernan, Bethan, and Harry Davies. "'The Machine Did It Coldly': Israel Used AI to Identify 37,000 Hamas Targets." *The Guardian*, April 3, 2024.

Davies, Harry, Bethan McKernan, and Dan Sabbagh. "'The Gospel': How Israel Uses AI to Select Bombing Targets in Gaza." *The Guardian*, December 2023.

Creyghton, Roos. "The Lavender Precedent." Action on Armed Violence (AOAV), November 6, 2025.

Sariel, Yossi. *The Human-Machine Team.* 2021.

Congressional Research Service. "Department of Defense Counter-Unmanned Aircraft Systems." CRS Report R47886, 2026.

Chapters 10 and 11

Defense Intelligence Agency. "Annual Threat Assessment." 2025 and 2026 editions.

Shah, Raj M., and Christopher Kirchhoff. *Unit X: How the Pentagon and Silicon Valley Are Transforming the Future of War.* Scribner, 2024.

Chapter 12

PitchBook. "Defense Tech Funding Data." 2025-2026.

Government Accountability Office. "F-35 Joint Strike Fighter: Program Reports." 2024.

Chapters 13 and 14

Human Rights Watch. *Under the Gun: How Autonomous Weapons Undermine Six Core Human Rights.* 2024.

International Committee of the Red Cross. "ICRC Position on Autonomous Weapon Systems." Updated May 2021.

Reaching Critical Will. "GGE on LAWS: March 2026 Session Report." March 11, 2026.

Kasparov, Garry. *Deep Thinking: Where Machine Intelligence Ends and Human Creativity Begins.* PublicAffairs, 2017.

Scharre, Paul. *Four Battlegrounds: Power in the Age of Artificial Intelligence.* W.W. Norton, 2023.

General References

Royal United Services Institute. Drone warfare analysis series, 2022-2026.

United Nations General Assembly. Resolution 79/62, "Lethal Autonomous Weapons Systems." December 2, 2024.

US Department of Defense. Directive 3000.09, "Autonomy in Weapon Systems." Updated January 25, 2023.

Arms Control Association. "The Missile Technology Control Regime at a Glance." Updated 2025.

Singer, P.W. *Wired for War: The Robotics Revolution and Conflict in the 21st Century.* Penguin, 2009.

Singer, P.W., and August Cole. *Ghost Fleet: A Novel of the Next World War.* Mariner Books, 2015.

Also by C. Scott Lannon

The Ghostwriter Machine

How to build an AI-powered publishing business from scratch. The definitive playbook for the creator economy's next evolution.

The Sleeper Agent Series

A Silicon Valley engineer discovers her AI creation is already three steps ahead. As the line between tool and threat dissolves, the people who built the most powerful technology in history realize they may no longer control it. AI thriller fiction for readers who want their tech plausible and their stakes real.

The Retrieval Program

A Senate staffer stumbles onto evidence of a recovery program that was never supposed to exist. Ten books follow the conspiracy from the halls of Congress to underground military installations to first contact. UFO mystery thriller fiction for readers of Blake Crouch and Michael Crichton.

High Ticket

A dark comedy about the online guru economy and the marks who fund it. Standalone novel.

Root Access

A tech noir mystery where the code is the crime scene. Standalone novel.

Visit cscottlannon.com/books for new releases and exclusive content.

Stay Informed

Want to stay current on the autonomous weapons revolution?

Join C. Scott Lannon's reader list:

https://www.cscottlannon.com/books/series/hunter-killer

You'll get:

A free copy of "The Five Drones That Changed Warfare" — five systems mapped onto the Hunter-Killer spectrum, from the $200 FPV drone to the autonomous swarm

Monthly intelligence briefs on drone warfare, defense technology, and autonomous weapons policy

First access to new books before public release

One email per month. No spam. Unsubscribe anytime.

Coming Next: The $200 War

The Hunter, Killer series continues...

The math is simple. A drone costs less than a bicycle. A tank costs more than a house. The drone wins.

That single economic fact — the most extreme cost inversion in the history of warfare — is rewriting the rules of conflict faster than any government, any military, any alliance can adapt. Nations that spent decades building billion-dollar fighter jets and aircraft carriers are watching $200 weapons neutralize them from a basement apartment. Countries that could never afford a modern air force now have one. Groups that were never supposed to have precision strike capability are assembling it on a folding table.

The first book in this series told you how the autonomous weapons revolution works. The next one tells you what it costs — and who pays.

The $200 War takes the economics of drone warfare and follows the money: from the Ukrainian startup founders building weapons from consumer electronics, to the Pentagon procurement officers watching their budgets become obsolete, to the arms dealers in Shenzhen who will sell autonomous strike capability to anyone with a wire transfer. The most important weapon of the twenty-first century costs less than dinner for two. The implications are just beginning to land.

The $200 War — Hunter, Killer Book 2 Coming Soon

https://www.cscottlannon.com/books/series/hunter-killer — Get notified the moment it releases.

Made in United States
Cleveland, OH
04 May 2026

36590909R00157